Simulation-Based Case Studies in Logistics

T0142115

Yuri Merkuryev • Galina Merkuryeva
Miquel Àngel Piera • Antoni Guasch
Editors

Simulation-Based Case Studies in Logistics

Education and Applied Research

 Springer

Yuri Merkuryev, Dr. habil. sc. ing
Galina Merkuryeva, PhD, DSc

Department of Modelling and Simulation
Riga Technical University
Kalku Street 1
LV-1658 Riga
Latvia

Antoni Guasch, PhD
Escola Tècnica Superior d'Enginyeria
Industrial i Aeronàutica de Terrassa
(ETSEIAT)
Universitat Politècnica de Catalunya
Planta 2, C/Gran Capità, 2–4
08034 Barcelona
Spain

Miquel Àngel Piera, PhD
Escola Técnica Superior d'Enginyeria
Departamento de Telecomunicació e
Ingeniería de Sistemas
Universidad Autónoma de Barcelona
Edifici Q, Campus de la UAB, Bellaterra
08193 Barcelona
Spain

ISBN 978-1-84996-825-6   e-ISBN 978-1-84882-187-3

DOI 10.1007/978-1-84882-187-3

A catalogue record for this book is available from the British Library

Cover design: eStudio Calamar S.L., Girona, Spain

Printed on acid-free paper

9 8 7 6 5 4 3 2 1

springer.com

Preface

Wide global market industry competition and customer product quality requirements are key factors that have forced engineers and economists to improve their operational decisions by integrating production, transport and service operations.

A common logistics perspective dealing with a proper coordination of all material movements and processing activities drives the reader through different chapters of this book. The specific application areas, in which simulation techniques are applied, demonstrate that improving key performance indicators of a real system requires not only addressing its technical aspects, but also designing tactical and operating procedures that would provide both the operational efficiency and economical practicality.

Simulation models have proved to be useful for examining the performance of different system configurations and/or alternative operating procedures for complex logistic and manufacturing systems. It is widely acknowledged that simulation is a powerful computer-based tool enabling decision-makers in business and industry to improve their organisational and operational efficiency. However, several limitations appear when trying to find a feasible solution to a logistic problem as only a limited number of simulation scenarios can be evaluated within acceptable time constraints.

The book is intended for intensive learning about the application of simulation as a decision support tool to tackle complex logistic problems. Case studies in the book are intended to allow the reader to follow and integrate typical phases and activities of a simulation-based study that lead to problem solving. A short list of such typical phases includes: problem formulation and setting of objectives; model conceptualisation; data acquisition and formalisation; simulation model development, verification and validation; experimentation; analysis of simulation output and making conclusions.

A key aspect to succeeding with the use of simulation techniques is the modelling activities. It should be noted that while most industrial operational procedures are based on the extension of current and past operating practices, development of simulation models could support defining new operational procedures at a fundamental level. Simulation allows identifying the role and operating methods of all members that interact during operational activities, as well as understanding the propagation of consequences of any potential decision, in order to deal with a safe and economically viable system.

There are different methodologies that have been used traditionally to develop simulation models in different areas, but modelling of logistic systems cannot be considered as a pure science. Representation of a logistic system depends on the experience of the modeller to identify a proper abstraction level at which system dynamics should be described, a formalism to be used in order to specify the system, and the clarity to discern between what is important and what can be neglected in the model to satisfy the goal of simulation experiments.

The book describes and illustrates different approaches to developing simulation models at the right abstraction level to be used efficiently by engineers when dealing with strategic, tactical or operational decisions in logistic systems. The book presents 12 simulation-based case studies based on results of the applied research performed by the authors.

These case studies cover a wide range of topics under a common objective, i.e. providing decision support for increasingly complex problems in the logistic area. They address core characteristics of typical logistic problems which can have different characteristics viewed from different perspectives.

While the case studies in this book share some commonality, they certainly make unique contributions in the following three main areas:

Manufacturing and Service Systems:

- *Manufacturing System Planning and Scheduling,* by Merkuryeva and Shires, tackles a very challenging subject regarding the use of simulation models for tuning quickly, and at a very low cost, production schedulers to find optimal configurations of their rules and parameters. Modular simulation models of the entire business/manufacturing system and a production anodising stage sub-model are developed in the ProModel software in order to test off-line effects of various scheduler configurations, avoiding disturbance of a real production process.
- *Hospital Resource Management,* by Aguilar, Castilla and Muñoz, proposes a hospital management tool to improve hospital efficiency by using a simulation model as a key source to obtaining a deeper knowledge on logistic processes and supporting decision making on resource redistribution. The Java discrete-event simulation system SIGHOS is developed and used to analyse different scenarios, providing a better resource distribution according to a priori knowledge of effects that management decisions would have throughout the hospital.
- *Flexible Manufacturing Systems,* by Piera, Narciso and Buil, illustrates advantages of using the coloured Petri net formalism to specify conceptual models of flexible manufacturing systems. The authors pay special attention to explaining how to develop a decision support system that evaluates the whole search space to tackle true flexibility of production systems by means of simulation.
- *Warehouse Order Picking Process,* by Merkuryev, Merkuryeva and Burinskiene, provides an MS Excel-based simulation model developed in order to analyse the influence of routing methods on picker travel distance in a wide-aisle warehouse. The picking process is a critical supply chain component for many companies. Proper warehouse configuration, storage policy, tray replenishment policy, and

other factors are important not only to reduce the delivery time, but also increase productivity while maintaining quality factors at competitive costs. This chapter focuses on a challenging simulation-based optimisation problem of finding appropriate routing methods to minimise the picker travel distance.

Transport Systems:

- *Factory Railway System,* by Guasch, Figueras and Fonseca, focuses explicitly on the analysis of a factory railway system using a simulation model to identify current limitations and potential infrastructure and resource investments to cope with a major increase in production. The conceptual model is formalised in the coloured Petri net formalism and the simulation model is developed in Arena©.
- *Material Handling System,* by Neumann, introduces a simple but efficient model to analyse the performance of a material handling system and to understand the load limit of a real system that consists of a warehouse, production and order-picking areas, and to analyse its ability to cope with a future load. The problem is characterised by numerous crossing flows of palletised raw materials, products and packaging material. The conceptual model is developed in the DOSIMIS-3 simulation package.
- *Vessel Traffic in the Strait of Istanbul,* by Uluşçu, Özbaş, Altiok, Or and Almaz, describes experiences of the authors in the decision-making area by modelling the complexity of operations in the Strait of Istanbul. The simulation model is developed in Arena© and incorporates an algorithm to schedule vessel entrances to the strait. The strait traffic rules and regulations, and transit vessel profiles, along with local traffic and other vessels, pilotage and tugboat services, and meteorological and geographical conditions are modelled, thus providing a tool to analyse policies and decisions regarding management of traffic, risks and vessel delays.
- *Airport Logistics Operations,* by Piera, Robayna and Ramos, introduces a discrete-event system approach to describe the main actors that operate in an airport. Illustrative examples of short-term solutions to mitigate delay propagation in Palma de Mallorca Airport are presented. An Arena© simulation model, describing the main airport operations, demonstrates benefits and handicaps of oversizing pushback resources with respect to improving collaborative decisions.

Supply Chain:

- *Supply Chain Dynamics,* by Hennet, examines the influence of different policies on management of a virtual enterprise in order to satisfy consumers' needs in the most efficient and profitable way, while avoiding the well-known 'bullwhip effect'. An algebraic model is introduced that allows one to compare production and ordering policies such as an inventory-based policy, an order-based policy and a mean demand-driven policy.

- *Pharmaceutical Distribution Network,* by Van Landeghem, tackles a challenging problem of optimising transportation modes in a distribution network of pharmaceutical goods, where delivery times are critical quality factors, and transport savings compete with the cost of opening and running warehouses.
- *Supply Chain Cyclic Planning and Optimisation,* by Merkuryeva and Napalkova, tackles a very challenging multi-objective stochastic optimisation problem: multi-echelon supply chain planning. It is characterised by a large number of decision variables and conflicting objectives. Several simulation optimisation scenarios are introduced in order to analyse and compare abilities of different optimisation methods and tools. In particular, the SimRunner® and OptQuest® add-on optimisation software and a hybrid simulation optimisation algorithm and tool introduced by the authors illustrate experimentation scenarios under specific cyclical constraints.
- Finally, *Fresh-Food Supply Chain,* by Bruzzone, Massei and Bocca, tackles a difficult problem of modelling fresh-food supply chains considering all the inter-related constraints and variables: time-to-market, traceability, transport/storage conditions, handling, production/process control, demand variability and seasonal behaviours.

Riga, Latvia – Barcelona, Spain   *Yuri Merkuryev*
August 2008   *Galina Merkuryeva*
                                                                                 Miquel Àngel Piera
                                                                                 Antoni Guasch

Acknowledgements

This case study book owes its appearance to the simulation community: academicians, researchers and industrial users that are permanently contributing to extending the use of simulation as an efficient tool to improve performance of complex logistic systems.

The origins of the book are traced to a joint work meeting of the McLeod Institute of Simulation Sciences (MISS, http://www.simulationscience.org) that took place at the Universitat Autònoma de Barcelona. We sincerely appreciate many public and private institutions and organisations that have indirectly contributed to supporting the scientific collaboration between the editors and authors. Special thanks to Riga Technical University, Universitat Politécnica de Catalunya, Universitat Autònoma de Barcelona, the International Mediterranean and Latin America Council of Simulation (I_M_CS, http://www.i-m-cs.org), and The Society for Modeling and Simulation International (SCS, http://www.scs.org), which supports many activities in the simulation area worldwide. In particular we express our recognition of the efforts of the organisers and participants of the annual International Mediterranean Modelling Multiconference (I3M, http://www.liophant.org/i3m/), which creates a real framework to exchange information, knowledge and experience acquired by top experts in the areas of logistics and simulation.

The editors would like to thank the anonymous reviewers of the book; it has greatly benefited from their very valued comments and suggestions.

We are very grateful to Olesya Vecherinska, Liana Napalkova and Tatyana Lagzdina of Riga Technical University for their accurate typing and formatting with numerous revisions.

Finally, let us wish that this book will become the seed for a series of case study books in the simulation area.

Contents

Contributors

Rosa María Aguilar Chinea
Dpto. Ingenieria de Sistema y Automatica, Facultad de Física y Matemáticas, Universidad de La Laguna, Calle Astrofísico Francisco Sánchez SN, 38271 La Laguna, Tenerife, Spain

Özhan Alper Almaz
Department of Industrial and Systems Engineering, Rutgers University, 96 Frelinghuysen Road, Piscataway 08854 NJ, USA

Tayfur Altiok
Laboratory for Port Security, Department of Industrial and Systems Engineering, Rutgers University, 96 Frelinghuysen Rd., Piscataway NJ 08854, USA

Enrico Bocca
Simulation division, MAST S.r.l., via Magliotto 2, 17100 Savona, Italy

Agostino Bruzzone
MISS-DIPTEM, University of Genoa, Via Opera Pia 15, 16145 Genova, Italy

Roman Buil Giné
Department of Telecommunication and System Engineering, High and Technical School of Engineering – ETSE, Autonomous University of Barcelona (UAB), Q building, UAB Campus, 08193 Bellaterra (Cerdanyola del Vallès), Catalonia, Spain

Aurelija Burinskiene
Department of International Economics and Business Management, Faculty of Business Management, Vilnius Gediminas Technical University, Sauletekio av. 11, LT-10223 Vilnius, Lithuania

Iván Castilla Rodríguez
Dpto. Ingenieria de Sistema y Automatica, Facultad de Física y Matemáticas, Universidad de La Laguna, Calle Astrofísico Francisco Sánchez SN, 38271 La Laguna, Tenerife, Spain

Jaume Figueras i Jové
Automatic Control Department, Universitat Politècnica de Catalunya, Rambla Sant
Nebridi 10, 08222 Terrassa, Spain

Pau Fonseca i Casas
Department of Statistics and Operations Research, Universitat Politècnica de Catalunya, Jordi Girona 1–3, Campus Nord, Ed C5 (LCFIB), 08034 Barcelona, Spain

Antoni Guasch
Automatic Control Department, Universitat Politècnica de Catalunya, Rambla Sant
Nebridi 10, 08222 Terrassa, Spain

Jean-Claude Hennet
LSIS-CNRS, Faculte Saint Jerome, Universite Paul Cezanne, Avenue Escadrille
Normandie Niemen, 13397 Marseille Cedex 20, France

Marina Massei
Liophant Simulation Institute, University of Genoa, via Cadorna 2, 17100 Savona,
Italy

Yuri Merkuryev
Department of Modelling and Simulation, Riga Technical University, 1 Kalku
Street, LV-1658 Riga, Latvia

Galina Merkuryeva
Department of Modelling and Simulation, Riga Technical University, 1 Kalku
Street, LV-1658 Riga, Latvia

Roberto Carlos Muñoz González
Dpto. Ingenieria de Sistema y Automatica, Facultad de Física y Matemáticas, Uni-
versidad de La Laguna, Calle Astrofísico Francisco Sánchez SN, 38271 La Laguna,
Tenerife, Spain

Liana Napalkova
Department of Modelling and Simulation, Institute of Information Technology,
Riga Technical University, 1 Kalku Street, LV-1658 Riga, Latvia

Mercedes Elizabeth Narciso Farias
Escuela Técnica Superior de Ingeniería, Departamento de Telecomunicación e Ing-
enieria de Sistemas, Unidad de Ingeniería de Sistemas y Automática, Universidad
Autónoma de Barcelona, Edificio Q, piso 1, Despacho QC/1051, 08193 Bellaterra
(Cerdanyola del Vallès), Barcelona, Spain

Gaby Neumann
Institute of Logistics and Material Handling Systems, Otto-von-Guericke University of Magdeburg, P.O. Box 4120, 39016 Magdeburg, Germany

İlhan Or
Industrial Engineering Deptartment, Bogazici University, Bebek, 34342 Istanbul, Turkey

Birnur Özbaş
Industrial Engineering Department, Bogazici University, Gunay Kampus, Muhendislik Binasi, 34342 Bebek, Istanbul, Turkey

Miquel Àngel Piera
Logisim Research Group, Department of Telecommunication and System Engineering, Universitat Autònoma de Barcelona, ETSE Campus de Bellaterra, 08193 Cerdanyola del Valles (Barcelona), Catalonia, Spain

Juan José Ramos
Logisim Research Group, Department of Telecommunication and System Engineering, Universitat Autònoma de Barcelona, ETSE Campus de Bellaterra, 08193 Cerdanyola del Valles (Barcelona), Spain

Ernesto Robayna Fernandez
Airport Production Devision, Palma de Mallorca Airport, AENA, AENA I Building, 07611 Palma de Mallorca, Spain

Nigel Shires
Preactor International Ltd., Cornbrash Park, Bumpers Way, Chippenham, Wiltshire, SN14 6RA, UK

Özgecan Uluşçu
C1 Consulting, 129 Summit Ave., Suite 200, Summit NJ 07901, USA

Hendrik Van Landeghem
Industrial Management Department, Ghent University, Technologiepark 903, B-9052 Gent, Belgium

Abbreviations

ADS-B	Automatic dependent surveillance – broadcast
AMPL	A mathematical programming language
ANOVA	Analysis of variance
AO	Aircraft operator
ATC	Air traffic controller
ATM	Air traffic management
CDM	Collaborative decision making
CNC	Computer numerical control
CODESNET	Collaborative demand and supply networks
CPN	Coloured Petri net
CPU	Central processing unit
DC	Distribution centre
DES	Discrete-event simulation
DES	Discrete-event system
EDC	European distribution centre
EDI	Electronic data interchange
EEG	Electroencephalography
EOBT	Estimated out-block time
ERP	Enterprise resource planning
FCFS	First come first served
FIFO	First in first out
FMS	Flexible manufacturing systems
FTL	Full truck load
GA	Genetic algorithm
GPS	Global positioning system
ICU	Intensive care unit
KPI	Key performance indicator
M&S	Modelling and simulation
MAP	Modified atmosphere packaging
MARLIN	Models for advanced reorganisation in logistics of ichthyic nourishment
MES	Manufacturing execution system
MILP	Mixed-integer linear programming
MRP	Materials requirements planning
OBT	Out-block time

PACU	Post-anaesthesia care unit
PC Enquiries	Personal care enquiries
PDF	Probability density function
PH Enquiries	Pharmaceutical enquiries
PMA	Palma de Mallorca Airport
PN	Petri net
RSM	Response surface-based method
RTI	Real-time information
SKU	Stock-keeping unit
SLA	Service level agreements
SME	Small and medium enterprise
SP	Sailing plans
SQF	Service quality factor
TMA	Terminal manoeuvring area
TOT	Take-off time
V&V	Verification and validation
VIP	Very important person
VTS	Vehicle tracking system

Chapter 1
Factory Railway System

A. Guasch, J. Figueras and P. Fonseca

Abstract In this study, a coloured Petri net conceptual model and an Arena© simulation model were developed for analysing the railway flow of hot steel coils in a steel factory. The simulation goal was to analyse a number of flow and storage management policies in order to identify which scenario reduces the total number of mobile railway resources needed for the internal transportation of the coils. This study was part of a larger study whose main objective was to redesign a factory railway system and a harbour steel terminal, in order to cope with a considerable increase in steel production.

1.1 Introduction

In this case study we propose analysing a factory railway system's ability to cope with a major increase in production. The factory produces flat steel from hot coils. The managers aim to increase production from 1.2 million to 1.8 million tons per year, comprising 351 consecutive working days. Managers are concerned about the capacity of the railway infrastructure to respond to the new demands. A simulation study is proposed to identify current limitations and potential investments.

This case study is a simplification of a more extensive study. Neither the supply chain nor the railway resources have been included, and other, minor aspects have been simplified, since our main aim is to discuss the methodology. The study is confined to the analysis of the flow of hot steel coils in the factory, because the amount of railway resources needed depends heavily on the flow management policy.

Antoni Guasch, Jaume Figueras and Pau Fonseca
Universitat Politècnica de Catalunya, Spain
toni.guasch@upc.edu, jaume.figueras@upc.edu, pau@fib.upc.edu

Y. Merkuryev et al. (eds.), *Simulation-Based Case Studies in Logistics*
© Springer 2009

1.2   Aims of the Study

Our main aim is to propose improvements to the current logistics of the railway system in order to minimise the amount of railway resources needed for transport operations. In the original project, a second aim was to propose improvements or new investments in the railway infrastructure and in the harbour terminal. Since this goal lies outside the scope of this case study, we will assume that the railway system will be able to transport the required transportation orders.

The original study was executed in two steps. First, the initial operating conditions were modelled and simulated. The model was verified and validated using real data. Because the factory managers were involved in the project, they became more confident about the validity of the simulation approach. Second, the future load was included in the model and the proposed improvements were tested. In this case study we generated synthetic data on expected future production levels.

1.3   Description of the System

In this section we describe the system and the data used in the simulation. Although the data acquisition stage is not described here, we should not ignore the difficulties and the amount of time needed for this stage, which may take up to 40% of the time required for a simulation project.

1.3.1   The Factory

Figure 1.1 shows the system, the hot-coil storage areas and the internal railway system used to transport the hot coils.

The hot coils arrive at the factory via two different channels:
1. *By the external, national railway.* The hot coils are stored directly in the A2 storage area.
2. *By ship.* The capacity of the A1 storage area located in the harbour is limited. The hot coils that cannot be stored there must be moved to the A2 area using the internal railway system.

There are three storage areas for the hot coils:
1. *A1*, which is located in the harbour and has a maximum capacity of 20,000 t.
2. *A2*, which is located inside the factory and has a maximum capacity of 150,000 t.
3. *A3*, which is used to feed the factory processing line with hot coils and has a maximum capacity of 8,000 t. In the initial configuration, A3 is not used as a storage area.

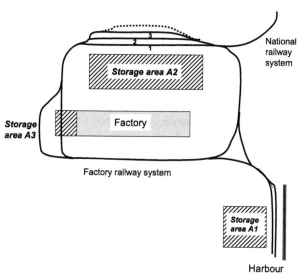

Fig. 1.1 The factory and its railway system

Since the capacity of A1 is limited, the current policy is to fill A1 completely and then move to A2 the coils that cannot be stored there. At the end, all the coils stored in A1 or A2 must be moved to A3 for processing. The candidate coils to be scheduled are selected from the storage areas. Thus, arriving coils must be stored before being designated as processing candidates. From the logistic point of view, important non-productive handling operations could be reduced by feeding A3 directly from the harbour; however, the coils' processing order and the capacity of A3 constrains this approach.

1.3.2 Arrivals of Hot Coils

The average weight of the coils is 21.7 t. The weight of the coils can vary significantly depending on their diameter and width; however, in this case study we assume that all the coils are of the average weight.

The hot coils received at the factory come from one of two suppliers:
- *Supplier S1*. Most of the steel is transported by shuttle train, although a minor part is transported by ship.
- *Supplier S2*. The steel is transported by ship.

The number of tons that the factory will receive from the two suppliers is:
- From supplier S1 by shuttle train: 700,000 t
- From supplier S1 by ship: 281,000 t
- From supplier S2 by ship: 819,000 t

1.3.2.1 Shuttle Train Arrivals from Supplier S1

There is an external national railway shuttle train on 95% of 351 working days. Since we are only interested in the daily flow of hot coils within the factory, the time of arrival is not needed.

The average transport weight of every shuttle is 2,103 t. A lognormal (2103, 320) distribution was obtained using a sample of the cargo weight of several trains.

Every steel coil has its own expected processing date. Ideally, the hot coils arrive at the factory in the order of their expected processing date. However, in practice, many coils arrive ahead of time and others arrive late. The normal (1.21, 14.20) distribution in days shown in Fig. 1.2 models the deviations with respect to the expected arrival time. The dot distribution of the figure is from the real sample.

Thus, the coils are processed according to the expected processing date, which is computed by taking the actual time at the moment of arrival and adding the time of the previous distribution to it. The processing order is also constrained by other factors, which are not considered here.

1.3.2.2 Ship Arrivals from Supplier S1

Ships from supplier S1 arrive on 19% of 351 working days. The cargo weight of each ship follows a uniform (2800, 5650) distribution. The delay distribution shown in Fig. 1.2 can also be applied to these coils to obtain the expected processing date.

1.3.2.3 Ship Arrivals from Supplier S2

Figure 1.3 shows the adjusted normal (3160, 583) distribution that models the cargo weight (in tons) of supplier S2 ships.

A ship arrives at the harbour on 74% of 351 working days. The lognormal (28.45, 6.86)-31.5 days distribution shown in Fig. 1.4 models expected deviations in the processing time with respect to the arrival time.

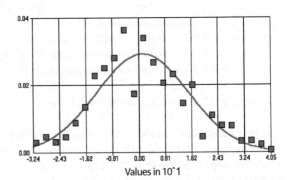

Fig. 1.2 Time arrival delay distribution for supplier S1 (days)

Values in 10^1

Fig. 1.3 Weight distribution of ships of supplier S2 (tons)

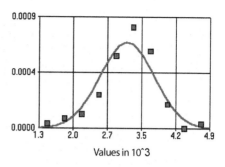

Values in 10^3

Fig. 1.4 Time arrival delay for supplier S2 (days)

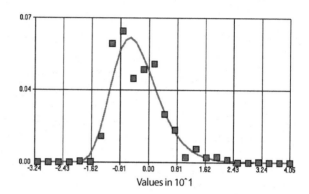

Values in 10^1

1.3.3 Hot-Coil Consumption

Pickling line is the first factory's processing line. A model of this line was obtained in a previous study [1]. The processing time (in minutes) is a function of the length l of the coil:

$$y = (0.1658 \times 10^{-5} \times l^2 - 0.003618) * 1.36 \tag{1.1}$$

Results obtained from a sample of 10,000 coils show that the coils' longitude follows a lognormal (719, 208) metres distribution. Furthermore, the failure model in the line is:

- Uptime: lognormal (578, 1060) minutes.
- Downtime: 86.3% of the failures follow a lognormal (15.42, 11.02) minutes distribution and 13.7% of the failures follow a lognormal (192, 150) minutes distribution.

Finally, every two weeks there is a maintenance stop lasting 16 hours. Thus, in the model, the line will stop for 16 hours every 14 days (these 16 hours count as part of the 14-day period). The previous processing time function was adjusted by a factor of 1.36, so that the production in a 351-day year would be 1.8 million t. This factor also accounts for additional non-linear aspects not included in the model.

We assume that every day at 10 a.m. the operations department updates the list of the next 10,000 t of coils that must enter the factory. For example, if at 10 a.m. on one day the processing queue is of 7,000 t, the operations department will select 3,000 additional tons to refill the processing queue to the 10,000-ton level. The coils have a processing order based on the expected processing time; however, once the coils are at the FIFO processing queue, the order cannot be modified. Thus, new coils arriving at the factory with an earlier expected processing time will not be inserted in the queue but will be added to the end of the queue. It is the logistics department's responsibility to issue transport orders during the day in order to ensure that the coils are available at the pickling line when they are needed.

1.3.4   Railway System and Storage Operations

The aim of the study is to propose improvements to the current logistics of the railway system in order to minimise the amount of railway resources needed for transport operations. Therefore, the main goal is to minimise the number of loaded trips that the locomotive has to make. We do not model the details of the transportation system or the loading/unloading operations in the harbour and storage areas.

1.3.4.1   Railway Operations

The locomotive transports 37 coils per trip (on average). There are two types of trips in the initial configuration:
- Trips of coils that cannot be stored in A1, from the harbour to storage area A2.
- Trips with coils scheduled for processing. Each trip starts at storage area A1, then the convoy is moved to A2 to complete the loading process. Finally, the loaded convoy is moved to A3 for processing.

1.3.4.2   Storage Operations in A3

A3 is able to hold two 37-coil convoys, both of which can be reached by the crane.

In the initial configuration, the coils were unloaded directly onto the pickling line's input processing buffer and the 8,000-ton capacity of the A3 storage area was not used. Thus, the convoys were used as input buffers on the pickling line. It was possible to use the convoys as a buffer at the 1.2 million-ton production level since the factory had enough wagons. However, since the number of available convoys could be a restriction if the production were increased to 1.8 million t per year, it would be advisable to unload the convoys in A3 as soon as possible to free them up for other transport operations. This strategy also imposes a significant constraint on the railway system. The coils in the convoy should be those that have to enter the pickling line according to the operations department's processing queue. This is the reason why the convoys have to be partly loaded in A1 and A2.

1.4 Modelling Methodology

Coloured Petri nets (CPNs) have proved to be successful tools for modelling logistics systems because of the conciseness with which they embody both the static structure and the dynamics of the system, the availability of mathematical analysis techniques, and their graphic nature [2–4]. Furthermore, CPNs are very suitable for modelling and visualising patterns of behaviour comprising concurrency, synchronisation and resource sharing, which are key factors when the aim is to optimise the performance of logistics systems.

Coloured Petri nets allow a high level of modelling by using colours that represent entity attributes of commercial simulation software packages. Both standard Petri nets [5] and CPNs [4] have been used extensively to evaluate the performance of production systems, because the model contains all the events and its interactions, together with the time consumed by each event.

The main characteristics of CPNs that provide a formalism appropriate for describing discrete-event simulation models are:

- All the events that might arise in each particular system state can be easily determined (coverability tree).
- All the events that might set off the firing of a particular event can be visually identified.

A modelling methodology that can support both of these characteristics for any type of discrete-event system is essential in tackling the performance improvement of a logistics system, from the conceptual model that describes all the event relationships to the codification of a simulation model that can support the decision task of optimisation routines at any stage in the evaluation process.

Other characteristics of CPNs enable this formalism to be used to specify logistics systems: they allow a logistics system to be specified at various levels of abstraction, depending on the modelling objectives; they allow a complex system to be specified using bottom-up techniques and advanced software engineering techniques, such as an iterative and incremental development process rather than a waterfall cycle and the promotion of a component-based architecture; and they can be considered to be graphical modelling tools that have few syntax rules.

Colour sets allow the modeller to specify the entity attributes. The output arc expressions allow the actions that must be coded in the event routines associated with each event (transition) to be specified. The input arc expressions allow the event pre-conditions to be specified. Finally, the state vector allows the modeller to understand why an event arises, and thus to introduce new pre-conditions (or remove them) in the model, or change variable and attribute values in the event routines to disable active events.

However, although CPN models contain the essential information required to build the simulation model, they are not widely used by commercial simulators as a mechanism for coding and specifying simulation models. For this reason, the model was coded using Arena© [6] (http://www.arenasimulation.com). The manual translation from CPN to Arena is fairly straightforward since there is a direct mapping between CPN structures and Arena's basic blocks.

1.5 Conceptual Model Building, Coding and Verification

In this section, we describe the conceptual model design, coding and verification steps that we followed. Readers who are interested in repeating the study should follow these steps.

The face validation technique was used to validate the model. It will not be described in detail here. This technique consists in discussing the conceptual model and the simulation results with experts to determine whether the model's behaviour is reasonable [7].

The model was specified using CPNs [2]. The following attributes (colours) were used in the model:

- c = integer; physical coil identifier
- id = integer; virtual coil identifier (in the database)
- s = {A1, A2, A3}; storage area where the coil is stored
- pl = pickling line
- rl = request for a scheduled load of coils
- $A1s$ = A1 storage area space
- ids = product $id*s$; combination of coil identification and the storage area
- $list$ = list(ids); list of ids
- $idlistA1$ = list(id); list of ids
- $idlistA2$ = list(id); list of ids
- $clist$ = list(c); list of cs
- e = extract order

1.5.1 Arrivals of Hot Coils

To code and check the arrivals of hot coils, follow the steps below.

a) Code the train arrivals from supplier S1. Check that approximately 700,000 t arrive over a period of 351 days. The number of coils that arrive is obtained by dividing the transport weight by 21.7 t per coil. Specify the expected processing date of each coil using the delay distributions.

b) Code the ship arrivals from supplier S1 and check that 281,410 t arrived in the 351-day period.

c) Code the ship arrivals from supplier S1 and check that 810,000 t arrived in the 351-day period.

1.5.2   Storage Areas and Unloading of Ships
and External Trains

The steps that must be followed are detailed below.

a) Initialise storage area A1 with 10,000 t of steel in coils (461 coils) and storage area A2 with 80,000 t of steel in coils (3,686 coils). The expected processing time is day 0 for all the initial coils. This aspect should be taken into account when one computes the warm-up period. Check the evolution of the stock of steel during the simulation period, which is the initial stock plus the arrivals minus the consumption at the pickling line.

b) Code and check the unloading of trains from supplier S1 in storage area A2. The following CPN (see Fig. 1.5) models the process. The arc expression nt`c is the used to represent the number of coils carried by the train on each trip; the colour c is an integer identifier that is unique to each coil; the colour id is also an integer identifier for coil records in the factory's databases; the place P3 stores information on all the hot steel coils stored in the factory. The thick arcs represent the flow of physical coils. Information such as the expected processing date is not represented in this CPN model.

c) Code and check the unloading of ships from supplier S1 and S2 in the harbour. The following CPN (see Fig. 1.6) models the process. A decision must be made as to whether the coil is moved to A1 (transition T11) or made available for transportation (transition T12) to A2. Transition T13 represents the grouping of 37 coils into a convoy. The current policy is to first fill A1 completely and then move the coils that cannot be stored there to A2.

A decision must be made as to whether the coil is moved to A1 (transition T11) or made available for transportation (transition T12) to A2. Transition T13 represents the grouping of 37 coils into a convoy. The current policy is to first fill A1 completely and then move the coils that cannot be stored there to A2.

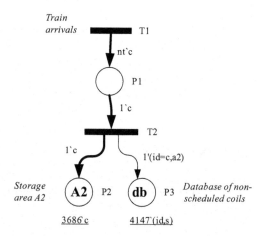

Fig. 1.5 CPN model of train arrivals

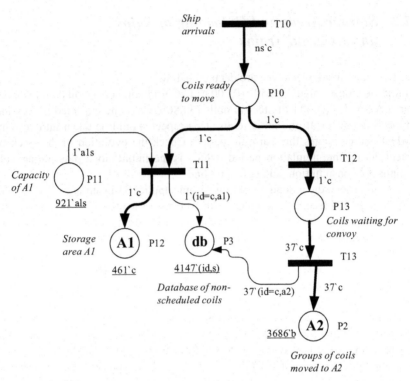

Fig. 1.6 CPN model of ship arrivals

1.5.3 Pickling Line

Next, the CPN (see Fig. 1.7) models the pickling line, the failure (T43) and maintenance (T45) stops, and the removal of the coil identifier (*id*) from the database
of scheduled coils (T42) when the corresponding physical coil is processed. T48
models a request for a new convoy load after the current load is processed.

To code the pickling line is needed to use the processing time function described,
the longitude distribution, the uptime and downtime distributions for failures and the
maintenance stops. Check that the line is able to process around 1.8 million t per year
(351 working days). Check the model's behaviour against the hypothesis that hot
coils will always be available. Transition T42 and T48 are not needed at this stage.

1.5.4 Scheduling

The next CPN (see Fig. 1.8) models the scheduling process and the grouping of
scheduled coils for transportation to A3 for processing. We assume that at 10 a.m.

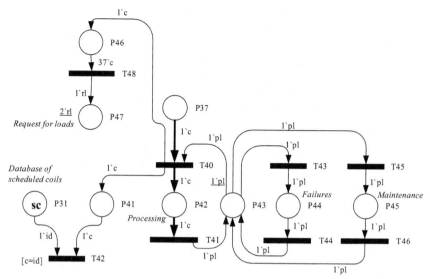

Fig. 1.7 CPN model of the pickling line

every day, the operations department updates the list of the next 10,000 t of coils that must enter the factory. For example, if at 10 a.m. on one day the processing queue comprises 7,000 t (n(P31) in the CPN), the operations department will select 3,000 additional tons (461-n(P31)) to refill the processing queue to the 10,000-ton level. The processing order of the coils in this simplified model is based on the expected processing time. However, once the coils are in the processing queue, the order cannot be modified. Thus, new coils arriving at the factory with an earlier processing time will not be inserted in the queue but will be added to the end of the queue. Virtual coils remain in P31 until the associated physical coil is processed in the pickling line.

Note that we need to model the physical coils that are stored, moved or processed in the factory and we also need a virtual coils (database entities) list, which is used to select the new coils that will be inserted in the processing queue. Coils are present in the database after being stored in A1 or A2; before that, they cannot be scheduled for production.

The scheduled coils are grouped in blocks of 37 coils (T32) to model the transportation grouping lists. Upon a request for a load (P53), the physical coils are removed from the storage areas and placed in P37. The T34 transition splits the list into a list that identifies coils stored in A1 and a list that identifies coils stored in A2. The guard function (*clist* = *idListA1*) indicates that the transition can be activated if the value of all the elements in *clist* is equal to the value of all the elements in idListA1. The sep(*clist*) function extracts all the individual coils from *clist*.

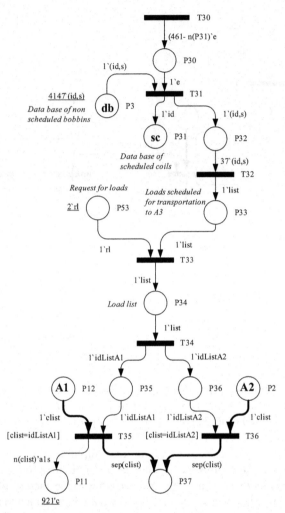

Fig. 1.8 Scheduling

1.6 Experimentation

This model does not include railway infrastructure. We assume that the railway system and the storage areas' infrastructure will be able to satisfy the transport orders. Using this model we can analyse different flow scenarios and identify the scenarios that may ensure the best performance.

The arrival patterns and pickling line yield are the same for all the scenarios. Figure 1.9 shows the number of tons that arrived and were consumed on 31 different days.

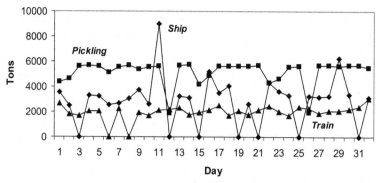

Fig. 1.9 Arrivals and pickling line consumption

For each scenario, we analyse:
- The number of trips that the internal locomotive has to make with loaded convoys.
- The stock in A1.
- The number of coil movement operations made by forklifts.
  1. Harbour: move a coil unloaded from a crane to A1.
  2. Harbour: move a coil unloaded from a crane to a convoy.
  3. A1: move a scheduled coil to a convoy.
  4. A2: move a coil from an external convoy to A2.
  5. A2: move a scheduled coil to a convoy.
  6. A2: move a coil from an internal convoy to A2.

1.6.1 Initial Scenario

This scenario is based on the factory working strategy before the investments. The hypotheses are:
- Fill A1 completely with steel arriving at the harbour and then move the coils that cannot be stored there to A2 in blocks of 37 coils.
- Scheduled coils are loaded onto convoys according to their processing order since the convoys are used as stock at the entrance to the pickling line, i.e. the pickling line is fed directly from the convoy. In this case, one group of scheduled coils is loaded in A1 and the other in A2.
- The storage capacity of A3 is not used.

The results of this initial scenario are shown in Table 1.1. The goal of this study is to propose new scenarios with the aim of minimising the number of loaded trips and the number of forklift operations. We are especially interested in the average number of daily operations since the factory will need resources for peak days.

Table 1.1 Initial scenario

	Loaded trips per day	Stock in A1	Daily forklift operations
Minimum	0	16,123	91
Maximum	24	19,985	1,140
Mean	15.2	19,333	563
Standard deviation	3.5	567.9	175

Figures 1.10 and 1.11 show a histogram of the number of loaded convoy trips per day and the stock in A1. Since with this policy the storage area is always nearly full, when two loaded ships arrive on the same day almost all the coils have to be moved to A2 by train. Thus, the factory will need enough railway mobile resources to cope with these peak days.

1.6.2 Second Scenario: Dampen the Harbour Arrival Peaks Using A1

One of the main problems of filling A1 first is that we cannot dampen the arrival peaks of raw steel since we are not allowed to leave the coils in the harbour overnight. These peaks appear when two ships arrive on the same day. On these days we need more railway mobile resources than average. In Fig. 1.12 we can see that in peak days we need to transport more than 6,000 t from the harbour to storage area A2.

This scenario tries to dampen the peaks by transporting a fixed number of non-scheduled convoys from the harbour to A2 if possible (if A1 is full we may need to transport more than four convoys). Thus, we force a fixed number of convoys even though there is still space available in the storage area A1. We have chosen four trips because there are on average 3.9 loaded trips from the harbour to A2 (other values might be tested as well).

The hypotheses of this scenario are:
- Four trips of non-scheduled coils from the harbour to A2 are made per day, if possible.

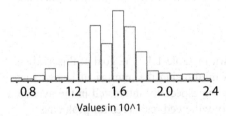

Fig. 1.10 Number of trips per day

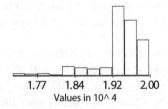

Fig. 1.11 Stock in A1

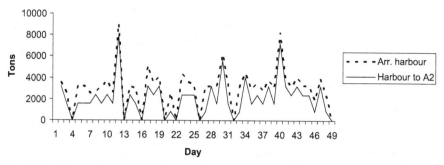

Fig. 1.12 Daily arrivals at the harbour and transportation to A2 (tons)

- Scheduled coils are loaded onto convoys according to their processing order since the convoys are used as stock at the entrance to the pickling line, i.e. the pickling line is fed directly from the convoy. In this case, one group of scheduled coils is loaded in A1 and the other in A2.
- The storage capacity of A3 is not used.

Though the average stock level in A1 is lower than it is in the initial scenario (Table 1.2), there is no significant reduction in the maximum number of daily operations. The average values are slightly higher because we are moving more coils than needed to A2.

1.6.3 Third Scenario: Dampen the Arrival Peaks and Use Storage Area A3 to Store 4,000 t of Scheduled Coils

The problems associated with using the wagons as stock in A3 are:
- The convoys are immobilised for a long period of time. This is not acceptable if resources are scarce.
- The coils in the convoy should be the ones that must be entered into the pickling line according to the scheduled list. This is the reason why the convoys have to be loaded partly in A1 and partly in A2. In this case, coils stored in A1 would need two trips to reach A3.

Table 1.2 Second scenario

	Loaded trips per day	Stock in A1	Daily forklift operations
Minimum	0	8,246	91
Maximum	21	19,898	934
Mean	15.9	14,230	578
Standard deviation	3.1	2,690	149

An improvement might be to reserve an area, within A3, to stock scheduled coils. In this case, convoys could be unloaded much more rapidly into A3 and complete convoys of scheduled coils could be fully loaded in A1 for destination A3 and in A2 for destination A3. In this scenario, the CPN has to be modified to model the use of A3.

In conclusion, the new hypotheses are:

- Four trips of non-scheduled coils from the harbour to A2 are made per day, if possible.
- Storage area A3 is used as a buffer for the pickling line. A3 has space reserved for three full convoys of scheduled coils arriving from A2 and two full convoys arriving from A1. In this case, convoys are not used as stock.
- The rest of the storage capacity of A3 is not used.

The results are shown in the Table 1.3. The maximum number of trips per day is 16, which is much lower than the 21 trips of the previous configuration and also lower than the 24 trips of the initial configuration. The mean number of trips is also lower.

Four convoys of steel are transported almost daily from the harbour to A2. This ensures that storage area A1 has enough capacity to store the arrival peaks in the harbour area (Fig. 1.13).

Table 1.3 Third scenario

	Loaded trips per day	Stock in A1	Daily forklift operations
Minimum	1	3,407	111
Maximum	16	19,985	1,080
Mean	9.2	12,623	576
Standard deviation	2.2	3,823	164

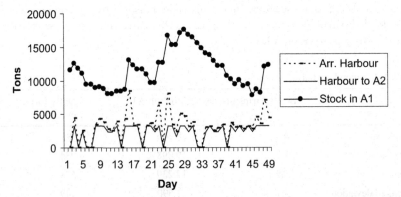

Fig. 1.13 Daily arrivals at the harbour, transportation to A2 and stock in A1 (tons)

1.6.4 Experimentation Overview

The current study was performed in three steps:

1. In the first step, a simplified analytical model was built. The results of the analysis of this simplified model gave us preliminary ideas about the behaviour of the system and the total amount of resources needed. Furthermore, the results were helpful in the simulation model validation phase.
2. In the second step, the flow simulation model described in this chapter was built to analyse the impact of several flow and storage management scenarios. It should be noted that the total amount of structural and mobile resources needed depend on the management policy selected. In this chapter, the third scenario is the one that shows the best performance since it reduces the peak transportation loads and the mean number of trips per day.
3. In the last step, the flow simulation model in the third scenario was expanded by including the railway's structural and mobile resources. The aim of this step was to determine the capacity of the available railway resources to cope with increases in production. This model took into account the travel times of locomotives and the time needed to load or unload convoys according to the number of forklifts available. The travel time depended on several factors, such as whether operations took place in the daytime or at night; whether they were carried out with or without a convoy; whether operations involved loading or unloading convoys; and whether they were uphill or downhill operations. As a result of this analysis, a new railway track was built in the harbour since there was a significant bottleneck in this area. Moreover, an automatic interlocking system was added to speed up train operations in the railway section next to storage area A2. As a result of the study, nine million euros were invested in the railway system.

1.7 Conclusions

The study presented here is a simplification of a real industrial railway transportation study. Using a simulation model, we analysed the logistics flow and storage management policy for steel coils in a factory, with the aim of minimising the use of resources.

The conceptual model was built using coloured Petri nets. CPNs are particularly suitable for modelling logistics systems since they are able to succinctly represent the concurrency, synchronisation and resource-sharing activities that are present in logistics and transportation systems.

Three different scenarios have been presented. The initial scenario represents the way the factory operated before the investment. It has been shown in this study that a 33% reduction in the number of loaded trips can be achieved (Scenario 3) by changing flow and storage management policies. This capacity margin eliminated the investing needs in new locomotive and mobile convoy resources.

1.8 Questions

1. Is it important whether the total number of tons arriving from each supplier is not exactly the number specified, or is it more important to model the irregular pattern of arrivals and tons for each shipment?
2. The failure model of the pickling line is based on historical data. Is it reasonable to expect that after the future investment the failure pattern will be the same?
3. If we did not have a model of the foreseen behaviour of the pickling line in the future, how would we model the raw materials consumption on the line?
4. Can this model be used to analyse the evolution of raw material stocks in the plant?
5. Let us assume that we know the expected processing time of the coils unloaded from a ship. What would be the impact of storing the coils with the earliest expected processing date in A1?
6. How could the remaining 4,000-ton space in A3 be used?
7. The behaviour of the railway system flow seems highly sensitive to the arrival patterns of trains and ships. What techniques could be applied to reduce the variability of the results?

References

[1] Guasch A, Piera MA, Figueras J (2004) Productivity analysis of a steel continuous pickling line using a hybrid model. In: Proceedings international workshop on harbour, maritime and multimodal logistics, HMS2004
[2] Jensen K (2007) Coloured Petri nets: basic concepts, analysis methods and practical use, vols 1–3. Springer-Verlag, Berlin
[3] Silva M, Valette R (1989) Petri nets and flexible manufacturing. Lecture notes in computer science, vol 424 (Adv Petri Nets 1989: 374–417)
[4] Zimmermann A, Dalkowski K, Hommel G (1996) A case study in modelling and performance evaluation of manufacturing systems using coloured Petri net. In: Proceedings 8th European simulation symposium, ESS '96, pp 282–286
[5] Piera MA, Guasch A (2001) Coloured Petri nets for simulation and FMS model maintenance. In: Proceedings EUROSIM'01
[6] Kelton WD, Sadowski RP, Sturrock DT (2004) Simulation with Arena, 3rd edn. McGraw-Hill, Boston
[7] Whitner RB, Balci O (1989) Guidelines for selecting and using simulation model verification techniques. In: Proceedings 21st winter simulation conference

Chapter 2
Manufacturing System Planning and Scheduling

G. Merkuryeva and N. Shires

Abstract This case study concerns support for customised solving of a production planning and scheduling problem in the piece-part medium-sized manufacturing company. To make the best use of an advanced scheduling tool and to find an optimal configuration of its rules and parameters, modular simulation models of the entire business/production process and production anodising stage are developed. Planning scenarios intended for optimising business processes in the company and different sequencing rules to improve processing of production orders are analysed. The improved approach and its benefits in practice are described.

2.1 Introduction

Modern production scheduling tools are very powerful and offer a vast range of options and parameters for adapting the tool's behaviour to the requirements of the real process. However, the more options exist, the more difficult it becomes to find the best configuration of the tool in practice. Even experts cannot often predict the effects of many possibilities. Testing out even a small number of possible configurations in reality and studying their effects on the real production process might take months and might severely reduce the overall performance. Hence, such tests are not feasible in practice. It is much faster, easier, safer and cheaper to test

Galina Merkuryeva
Riga Technical University, Latvia
gm@itl.rtu.lv

Nigel Shires
Preactor International Ltd., UK
nigel.shires@preactor.com

Y. Merkuryev et al. (eds.), *Simulation-Based Case Studies in Logistics*
© Springer 2009

and optimise a production scheduler using a simulation model than using the real process [1].

In order to make the best use of an advanced and sophisticated scheduling tool in the piece-part medium-sized manufacturing company and to find an optimal configuration of its rules and parameters, modular simulation models of the entire business/manufacturing system and production process anodising stage are built in order to test out the effects of various scheduler configurations [2]. Testing and optimisation of the scheduling tool configuration is carried out off-line by using simulation models. The real production process is not disturbed, and the optimal configuration can be found very quickly and at low cost.

2.2   Problem Formulation

Decorpart, a UK-based medium-sized manufacturer, produces a wide range of different small pressed aluminium parts in large quantities to a range of other consumer-focused businesses. Typical applications include spray assemblies for perfumes and dispenser units for asthma sufferers. The business lies in a highly competitive sector, and success depends on achieving high efficiency and low cost of manufacturing. Production scheduling is therefore very critical.

In the past, the company had already installed software tools supporting the scheduling of individual areas of the production process. To improve the overall company performance, increase its output and reduce the product lead time, they have planned to implement an automatic Preactor supply chain planning server – an overall scheduling system coordinating all local business and production areas. In order to deliver the best possible solution, the supplier of the scheduling tool, Pre-actor International (http://www.preactor.com) decided to use simulation for finding the optimal configuration of the scheduling tool.

The problem is to build a simulation tool, which will embrace the arrival of customer orders and sequencing of production orders to meet these demands. An important aspect is to model the production process itself in order to ensure that its main stages are optimally loaded at all times. The anodising stage is known to be particularly important for the overall production. Thus it has to be modelled in great detail and used in order to test to what extent the overall lead time of the orders can be reduced by optimisation of the anodising process stage.

The following key objectives are stated in this case study: (1) to model inter-related business and production processes at the company and to determine the overall lead time of orders, (2) to analyse and optimise business processes at the planning department dealing with processing of incoming enquiries and planning production orders, (3) to test the sensitivity of the overall production lead time to improvements, in particular, to determine whether introducing specific sequencing rules of production orders will decrease their total processing time at the anodising process stage.

Moreover, a simulation tool is aimed to be used for testing the configuration of the scheduling tool and for iterative optimising its performance off-line prior to its implementation and integration at the customer's site. The envisaged scheme is designed to complement and link together localised advisory systems previously installed on individual areas of the production process.

The main impact of simulation is expected to be a higher system throughput with lower product unit costs.

2.3 Modelling Approach

A custom-built business/manufacturing system model is created that simulates the arrival of orders, their queuing and their flow through all steps of the production process. For the overall coordination and schedule optimisation, each process stage is modelled as a group of machines with an overall capacity per day or per week. The model is built in a modular style so that each production stage could be further modelled to a greater level of detail. As mentioned above, the anodising process stage is known to be particularly important for the overall production. Thus this production stage is modelled in a greater level of detail following successful validation of the initial model.

Therefore the model of the anodising process is refined and the individual anodising tanks are described in detail, so that colour changeover and set-up operations could be studied more precisely. In this way, order queue ranking rules that minimise colour changes are introduced and tested as to what extent the overall lead time of orders can be reduced by optimisation of these rules at the anodising process stage.

Next, the Preactor scheduling tool is coupled with: (1) a high-level business/manufacturing system model, and (2) a detailed representation of the anodising process stage, both of which were developed using production simulation system ProModel [3] and used for finding the optimal configuration of the scheduling tool.

2.3.1 A High-Level Business/Manufacturing System Model

In this section we will provide the conceptualisation and input data analysis for a high-level business/manufacturing system model. It is aimed at modelling interrelated business and production processes at the company in order to analyse and optimise business processes at the planning department. These processes relate to the processing of incoming enquiries and planning of production orders confirmed by customers. The model is used to compare two alternative planning scenarios (see Sect. 2.5) and analyse the benefits of introducing an advanced production scheduling and capacity optimisation tool at the company with the maximal response time of 0.1 hour per enquiry.

Model conceptualisation. The custom-built entire business/manufacturing system conceptual model is given in Fig. 2.1. The model simulates the arrivals of enquiries and their processing time; generates orders becoming confirmed by customers and their planning time, and shows the queuing of the production orders for processing. There are two types of incoming enquiries – pharmaceutical enquiries and personal care enquiries, which are denoted as *PH_Enquiries* or *PC_Enquiries*, respectively.

Production itself consists of the following processing stages: pressing, degreasing, jigging, anodising and packing. In this model the production of orders does not need to be modelled in detail. So, in each production stage the individual machines are modelled as a group with an overall capacity per week. No queues are defined for locations used to simulate different production stages in the system model.

The following parameters could be controlled in the system: the number of planners that process enquires from customers as well as respond to customers and plan confirmed orders for production; the response time for enquiries, and planning time for confirmed orders. These system parameters define the controllable variables in the simulation model.

Parameters such as time between arrivals of enquiries, customer response time to confirm or cancel enquiries, the probability of an enquiry becoming confirmed or becoming an order, and order processing time for different production stages could not be controlled in the system. These parameters are regarded as environmental variables in the model.

The system key performance indicators such as total revenue, an average lead time, the percentage of cancelled enquiries and utilisation of planners define the model performance measures.

Data collection and analysis. Based on the analysis of the historical data and taking accounts, their stochastic nature probability distributions given in Table 2.1 are derived. For example, the time between arrivals of *PC_Enquiries* is exponentially

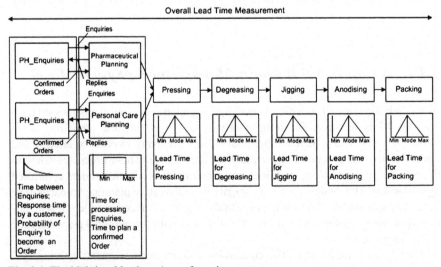

Fig. 2.1 The high-level business/manufacturing system

distributed with the mean equal to 20, and processing time of the enquiries is uniformly distributed with the mean and half range equal to 35 and 5, respectively (see ProModel distribution functions in [3]). These distributions are used in the model to generate the time between arrivals of enquiries, processing times of the enquiries, an average response time from a customer and actual planning time of confirmed orders. About 33% of all incoming enquiries are *PH_Enquiries*. The probability of enquiries becoming an order decreases as a function of the planning department response time including enquiries queuing time and is given in Table 2.2. On the other hand, the value of confirmed orders received by the company increases as a function of the planning response time. In the case study, the average order value is defined.

An average order lead time in each production stage is defined by the triangular distribution with the following parameters: min = 1,080, mode = 1,440 and max = 1,800.

Currently *PH_Enquiries* are processed by one planner, and *PC_Enquiries* are processed by another three planners that spend about 70% of their working time on planning operations. The working day is eight hours long starting from 9.00 a.m. Planning staff employment costs per year are fixed.

Model building. The entire business/manufacturing system simulation model is built using the ProModel basic modelling elements such as locations, entities, arrivals and processing. A number of variables are defined as well. Some of these variables are counters which record statistics about cancelled enquiries, orders in process, completed orders, etc. So-called processing variables are introduced to make it easier to change processing times in the model.

Visualisation of the model is presented in Fig. 2.2. On-line and off-line statistics are provided. Simulation outputs reflecting the model dynamics (i.e. *Waiting enqui-*

Table 2.1 Probability distributions (all values are given in minutes)

Data	Distribution type	Distribution
Time between arrivals of enquiries		
PH_Enquiries	Exponential	E(60)
PC_Enquiries	Exponential	E(20)
Processing time of enquiries	Uniform	U(35, 5)
Response time from a customer	Constant	24 * 60
Actual planning time of confirmed orders	Uniform	U(55, 5)

Table 2.2 Probability of enquiries becoming an order

Enquiries becoming confirmed (%)	Planning response time
50	< 1 hour
20	1–8 hours
10	24–48 hours

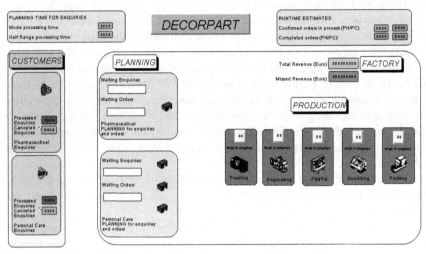

Fig. 2.2 A high-level business/manufacturing system model screenshot

ries, Completed orders, Total revenue) can be followed on the model main screen. Results of conducted experiments are automatically saved in the model database and presented in Excel spreadsheets.

In order to check if the model reflects the real process adequately, a set of historical data was compared with the data produced by the simulation model. It was found that the model and the real process produced more or less identical results.

2.3.2   A Low-Level Anodising Process Stage Sub-Model

Model conceptualisation. The low-level anodising process stage sub-model [4] is aimed at testing whether the implementation of specific sequencing rules of incoming production orders will decrease their total processing time at a batch anodising plant.

Batch anodising refers to anodising of series of small parts produced in batches. The anodising process contains the following steps. First, the metal parts are batched on racks. After batching the metal parts are degreased and cleaned. Then batches of cleaned metal parts are put in a bath of acid where the oxide film around the aluminium is created. After that the aluminium parts are rinsed with cold water. Then the oxide film around the aluminium is coloured with a spray. This spray, which is also called as a dye, is typically a kind of paint, mixed with water. Dying can be done in several steps in order to provide the right colour. Changing the colour of the dying process is a bottleneck in a real system. Coloured parts are rinsed first with cold water and then with hot water.

The model itself simulates the individual anodising tanks so that colour changeover, set-up operations and processing times can be modelled. Based on the historical data about order processing, the most probable list of incoming orders to be weekly processed is generated in the model. Specific sequencing rules of incoming orders are simulated and tested in order to decrease the total processing at the anodising stage. Production rate, which is defined as an average number of flight bars processed per hour, and the frames utilisation coefficient are used to measure the effectiveness of the anodising plant itself.

The anodising sub-model black-box diagram is presented in Fig. 2.3. The sequence numbers of incoming orders that have to be processed in a week is controlled in the model. The order quantity, part colour and used frame type for incoming orders are regarded as environmental or independent variables. If these properties are given, the other properties of orders in the order list can be determined. Other environmental variables are the number of frames in stock, the time it takes to load and unload flight bars, the time it takes to set-up flight bars between the processing of different colours and the processing time necessary to anodise one batch of components.

The most important performance indicator is defined as the total processing time of all orders in the order list. Among other performance indicators that could be used to control an anodising process in the real system, the following performance measures can be mentioned: average production rate, frame loading efficiency, flight bars utilisation and plant productivity.

Data collection and analysis. First, based on the analysis of historical data about the orders that were planned and processed at the plant in a certain period the general order list is created. It includes the following input data: week number, order number, order quantity, colour, frame type and frame capacity, the number of frames in stock, number of batches and sequence number (Table 2.3).

The last four digits of the order number, *Order no.*, refer to the code of the colour which the components should get. Each frame type has a different number of components that can be placed upon it, which is called as *Frame capacity*. The number of frames of a specific type available is called as *Frame in stock*. Only three frames can be loaded on each flight bar.

Processing time of one batch of the components in a flight bar depends on the program that is used in the anodising process is defined by a sequence number *Seq. no.* in Table 2.3. Based on the input data analysis, processing times are described by the triangular distribution and generated in the simulation model. For example, for sequence 8, which is used by orders with colour code 0001 the triangular distribution with endpoints (54, 72) and mode at 58 is used in the model.

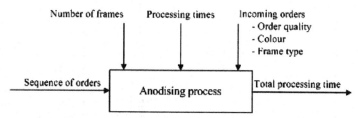

Fig. 2.3 Anodising sub-model black-box diagram

Table 2.3 A fragment of the general order list

No.	Week	Order no.			Order qty (× 1,000)	Colour		Frame Type	Frame capacity	Frames in stock	Number of batches	Seq. no.
1	1	1135	1	0001	100	Bright	Silver	C1	1292	15	26	8
2	1	1135	1	0134	100	Bright	Gold	C1	1292	15	26	6
3	1	1407	0	0003	2	Bright	Gold	D2	2400	11	1	6
4	1	1135	1	0134	55	Bright	Gold	C1	1292	15	15	6
5	1	0803	0	0058	25	Bright	Gold	D2	2400	11	4	8
6	1	1210	1	0001	300	Bright	Silver	L2	2500	18	40	8

Table 2.5 A fragment of the input order list

No.	Colour code	Qty (× 1,000)	Frame type	Frame capacity	Processing time (min)	Processing time (mode)	Processing time (max)	Batch no.	Frames no.	Frames left
1	0058	28	7	3456	54	58	72	3	9	0
2	0003	225	2	1900	64	87	92	45	135	0
3	0001	224	6	3456	54	58	72	22	65	2
4	0001	711	3	2400	54	58	72	99	297	0
5	0058	139	6	3456	54	58	72	14	41	2
6	0001	93	4	2500	54	58	72	13	38	2

Table 2.4 Empirical probability distribution for order quantity (colour number 0001)

From	To	Probability
0	100	0.407
100	200	0.507
200	300	0.759
300	500	0.815
500	600	0.928
600	700	0.963
700	800	0.981
800	1000	1

Second, based on the general order list the most probable list of incoming orders to be weekly processed in the model is generated. The number of orders in this order list is fixed equal to the average number of orders in a week. Frequencies of order colour and order quantity as well as of the frame type to be used are derived from the general order list data and defined by empirical distributions (see an example in Table 2.4). For simplification it is assumed that order quantity and frame type depends on the product colour to be anodised. Fitted probability distributions are used to generate the most probable list of orders or so-called input order list. A fragment of the completed input order list is given in Table 2.5.

Note that parameters of the probability distribution that fit processing times (such as minimum, maximum and most likely value), the number of batches that an order should be split up in, the number of frames necessary to process all batches and the number of frames left are also included in the Input order list. The Input order list is generated in Excel spreadsheets that allow automated retrieval data from it within the simulation experiments.

Model building. The anodising process stage sub-model is built using the Pro-Model basic elements and includes three types of locations: a location where entities that are batches in the model arrive, another location where processed entities move to and the number of locations where entities are being processed.

Figure 2.4 shows a screenshot of the model visualisation that is created by animation of pictures that simulates order arrivals and storage as well as colour change-over, set-up and order-processing operations. The user can follow the flow of batches from the arrival location and analyse the current stage of the anodising process for each order. Different colours are used for incoming and processed entities. Entities that are processed move on to the storage location.

On-line statistics are provided by three counters on the right-hand side of a screenshot that display the following performance characteristics of the anodising plant: the number of orders that are left to process, the number of batches left to process and the average number of processed batches per hour. Two additional counters along with the flight bars indicate the current number and the colour of the order that is currently being processed. Total processing time of all incoming

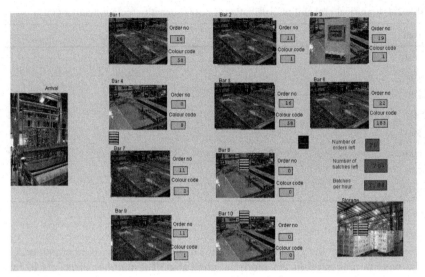

Fig. 2.4 The anodising process stage sub-model screenshot

orders, frames loading efficiency and plant utilisation can be found in the general simulation output report.

In the case study, validation of the anodising process stage sub-model is not described in detail. Note that similar to the entire business/manufacturing system model, in order to validate this model a set of historical data was compared with the data produced by the simulation model.

2.4 Experimentation

To identify the warm-up period, to select the replication length and the number of replications, and set these options in simulation experiments, we refer the reader to statistical methods of simulation output analysis and simulation options provided by ProModel simulation software [3].

2.4.1 Planning Scenarios for Business Process Optimisation

To understand the entire business/manufacturing model behaviour and define which input factors have important impacts on the model outputs, regression-type simulation metamodels were built in the case study. For example, the following regression equation was received, which shows the effects of input factors to *PC* order lead time in the system:

$$Lead\ time\ (PC) = 9277.03 - 21.05 * Enq + 4.83 * Ord + 0.62 * Enq^2 + 0.41 * Enq * Ord,$$

where *Enq* and *Ord* denote *PC_Enquiry* processing time and order planning time, respectively. As the result we conclude that the model outputs are more sensitive to enquiries processing time rather than to orders planning time.

Then to investigate how sensitive the model outputs are to the changes in the important inputs, these inputs were systematically changed and simulation outputs were observed. It was stated that if the response time for customer enquiries could be reduced by 5%, the total revenue of the company would grow by about 10%.

For business process optimisation within available system resources two optimal designs of the system using the ProModel SimRunner® Optimiser were generated. They define the optimal combination of enquiry processing time and order planning time that maximises the total revenue and minimises the lead time indicator, respectively. The results (see Table 2.6) show that the maximum revenue could be achieved if enquiry processing time does not exceed 6 minutes. This could be achieved by introducing the automatic Preactor supply chain planning server with a maximal response time of 0.1 hour, or 6 minutes per enquiry.

To improve the planning process at the company, two alternative scenarios were compared:
- *Scenario 1* in which the scheduling of individual areas of the production process is provided – the current situation with the maximal response time equal to 1 hour per enquiry, not including queuing time
- *Scenario 2* in which an overall scheduling system coordinates all local business and production processes – introducing the automatic Preactor supply chain planning server

The results of simulation experiments (Table 2.7) show that the number of cancelled orders in Scenario 2 can be decreased by 14–18%, which would cause the total revenue or the total value of confirmed orders to increase at least twice. This can be explained by a shorter enquiry processing time that provides a faster response to the customer and leads to a higher probability for enquiries to become an order.

Table 2.6 Comparison of two optimal designs

	Enquiry processing time (min, max)	Order planning time (min, max)	Revenue €	Leadtime, PH (min)	Leadtime, PC (min)
Maximised revenue	(4, 6)	(2, 8)	49,900,000	9,218.2	9,261.1
Minimised lead time	(1, 11)	(3, 7)	48,210,000	9,244.4	9,134.7

Table 2.7 Comparison of alternative planning scenarios

	Lead time (min)		Total revenue (€)	Cancelled enquiries (%)	
	PH	PC		PH	PC
Scenario 1	10,805	10,414	17,170,588.24	57	57
Scenario 2	9,793	9,617	41,758,823.53	43	39

Moreover, instead of four planners, only three would be needed if the new scheduling tool were introduced. Thus, employment cost can be saved as well.

Notice that the total revenue value was estimated based only on observations on the steady-state behaviour of the model. The counters for completed orders are stated for the replications including the model warm-up period. The last one is estimated almost by three weeks. The replication length is defined as twice as the warm-up period. While the planning department works only on weekdays, the production process continues 24 hours a day, seven days a week. After ten replications the variance in the output variable such as average lead time is small enough to get a half range of 5% average.

2.4.2 Testing Sequencing Rules for Processing Production Orders

The scheduling of order processing at a batch anodising stage is to be interpreted as a finite capacity scheduling problem. This is defined as the process of creating an operation schedule for a set of jobs that are to be produced on a limited set of resources. In the problem, the number of frames in a stock available for a specific frame type and the number of flight bars that the frames are loaded on are limited.

Since this frame type is limited, it could cause queues of orders waiting for free frames, while the flight bars could be empty. On the other hand, processing of production orders with different colours could lead to multiple set-up operations, while decreasing the number of necessary set-up operations will result in reducing the total lead time at the plant.

For testing different order sequencing rules four simulation scenarios were introduced in this case study (see Table 2.8). In Scenarios A0 and A1, single queue sequencing rules are applied. Scenario A0 represents the initial situation, in which the incoming orders are processed according to their arrival mode. In Scenario A1, the orders with the largest quantity of components are processed first. But in Scenario A2, the orders wait in separate queues determined by order colour and single sequencing rules are applied to orders within each queue. In Scenario A3, an order sequencing rule combination is used in which the colours that appear less frequently in the list are processed first, while within the group of the same colour, the orders with the largest number of components are processed first.

Table 2.8 Simulation scenarios

Scenario	Sequencing rules
A0	First-come, first-served
A1	Largest order quantity first
A2	Queuing by colour
A3	Less frequent colour first–largest order quantity combination

To implement sequencing rules for processing production orders in the simulation model, the input order list described in Sect. 2.3.2 was rescheduled in the way the scenarios describe. The difference between mean values of the total processing time of all incoming orders was estimated from simulation experiments for scenarios with specific sequencing rules and the initial scenario. The length of the simulation run was defined to be equal to the time between the start of the week, which represents the initial situation in the real system, and the time that all the week's orders had been processed. For each replication, the common random numbers were used to simulate both scenarios, leading to a lower variance of the mean estimate.

The results of simulation experiments with the detailed model of the anodising stage have demonstrated that introducing new specific sequencing rules for incoming orders could provide significant improvements. While comparing Scenario A0 and A1, 20 replications were performed for each scenario and the difference of two means $\mu_{A0} - \mu_{A1}$ was estimated as 11.51 hours with 95% confidence interval equal to (3.82, 19.9) hours (see Fig. 2.5, a). This led to the conclusion that the A1 sequencing rule for incoming orders in a week could reduce the total lead time of this stage by at least 4 hours, in some cases even by 19 hours. As a result, the production rate of the anodising stage will go up by 10%, and a significant increase in equipment utilisation and reduction of unit manufacturing cost can be achieved.

At the same time, the confidence interval for two other cases (see Fig. 2.5, b and c) contains zero. These results show that there is no significant difference between the mean total processing times produced by Scenario A0 and Scenarios A2 and/or A3, respectively, and there is no sufficient evidence to pick one alternative scenario over another one.

Then what-if analysis was performed to test whether the implementation of Scenario A1 is still an improvement if the number of frames in stock will be increased. In this case frames are not considered as limited resources in the real system. The results of comparison of sequencing rules with unlimited frames showed that Scenario A1 will not make a significant improvement compared to Scenario A0 (see Table 2.9).

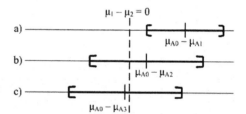

Fig. 2.5a–c Positions of the confidence intervals relative to zero

Table 2.9 Comparison of alternative sequencing rules with unlimited number of frames

Scenarios		Mean difference (hours)	95% confidence Interval	Significant
A0	A1	0.01	(−0.55, 0.58)	No
A0	A2	6.27	(5.85, 6.89)	Yes
A0	A3	6.23	(5.59, 6.86)	Yes

On the other hand, the orders queuing by colour in Scenario A2 will decrease the total processing time at least by 5.85 hours. At the same time, there will be no significant difference between Scenarios A2 and A3.

2.5 Conclusions

This case study demonstrates that the modular simulation models provide an inexpensive tool for an overall guidance and testing of advanced scheduling middle-scale software packages prior to their implementation at the customer's site.

The modelling approach used in the case study – to test and optimise advanced planning and control tools off-line by using simulation models rather than using the real process – can be applied to many other software tools, to higher-level (MRP; ERP tools) as well as to lower-level control tools (MES, warehouse control systems). On the other hand, the development of such relatively simple simulation tools in different industrial sectors could also provide an inexpensive approach to an overall guidance of small and medium-sized manufacturing towards the optimal conditions without resource to high-cost integration of expensive ERP systems and downstream control systems.

2.6 Questions

1. How can simulation help test and find the best configuration of the scheduling tool in a real system?
2. What is the range of scenarios for which simulation is used in planning and scheduling of the manufacturing system?
3. What is the main feature of the modelling approach applied in this case study?
4. What are the most significant differences between simulation models built within this approach?
5. What are the characteristics of the simulation model used for business process optimisation?
6. What are the characteristics of the simulation sub-model that is used for sequencing of the production orders at the anodising stage?
7. What does the confidence interval express about the order sequencing rules at the anodising stage?
8. Which techniques are used to validate the simulation models?
9. Define the main operational and financial benefits of this study.

Acknowledgment This work was supported by the SIM-SERV Thematic Network project 'Virtual Institute for Production Simulation Services' (http://www.sim-serv.com).

References

[1] Merkuryeva G, Shires N, Krauth J (2004) Simulation-based production scheduling and capacity optimisation in manufacturing SME's. In: Proceedings 11th international power electronics and motion control conference, vol 4, pp 225–230

[2] Merkuryeva G, Shires N (2004) SIM-SERV case study: simulation-based production scheduling and capacity optimisation. In: Proc 18th European simulation multiconference 'networked simulation and simulated networks', pp 327–333

[3] Harrell CR, Ghosh BK, Bowden RO (2004) Simulation using ProModel, 2nd edn. McGraw-Hill, New York

[4] Merkuryeva G, Shires N, Morrison R et al (2003) Simulation based scheduling for batch anodising processes. In: International workshop on harbour, maritime multimodal logistics modelling and simulation, pp 170–176

Chapter 3
Supply Chain Dynamics

J-C. Hennet

Abstract Supply chains show comparative advantages over traditional integrated manufacturing systems in terms of flexibility and quick adjustment to demand variations. However, the tendency to amplify demand variations upward the supply chain, known as the *bullwhip effect*, has been identified as a major drawback. This study proposes a modelling framework that allows for comparing several production and ordering policies: an inventory-based policy, an order-based policy and a mean-demand-driven policy.

3.1 Introduction

During the last fifteen years, supply chain analysis has become a major concern both in manufacturing theory and in industrial practice. In the extended sense, which is now prevailing in the literature, a supply chain associates all the enterprises that contribute to production and sale of a family of products (goods or services). In this view, an ideal supply chain can be seen as a virtual enterprise designed to satisfy some consumers' needs in the most efficient and profitable way. Several performance indices have been proposed to evaluate the quality of a supply chain, particularly in terms of costs and value, decisional integration, agility, reactivity and reliability. Some obstacles to performance maximisation have then appeared in the very nature of supply chains. In particular, the decisional autonomy of enterprises sets some limits to communication, coordination and integration between the interacting entities.

Jean-Claude Hennet
LSIS, Université Paul Cézanne, France
jean-claude.hennet@lsis.org

Y. Merkuryev et al. (eds.), *Simulation-Based Case Studies in Logistics*
© Springer 2009

Responsiveness and reactivity to demand changes may generate fluctuations in workload and inventory levels that may be amplified as they propagate upward along a supply chain. Since the work of H. Lee [1], such a phenomenon has been referred to as the *bullwhip effect*.

Observation of real industrial cases and simulation studies has shown examples of bullwhip effects. The bullwhip effect has been observed in many real supply chains, ranging from mechanical industries to several food sectors (see [2]). A case reported by the Supply Chain Simulation Workgroup of Santa Fe Institute Business Network concerns Procter and Gamble's Pampers division, who found 'huge swings in weekly demand and orders within their supply chain'.

Managers, researchers, engineers and students all over the world have been taught the bullwhip effect by playing the popular 'Beer Game', a simulation developed by John Sterman's group at MIT's Sloan School of Management [3]. The players of the game represent four nodes in an idealised supply chain for a Beer company: retailer, wholesaler, distributor and factory. Playing the game produces volatility much like that observed in real supply chains. As the backlog for orders increases, players order too many input products, forcing their suppliers into severe backlogs. Conversely, excessive production propagates downward along the supply chain and the decrease of orders amplifies.

Key factors for generating bullwhip effects are demand uncertainty and distributed lead times, but it has been shown that the loading and ordering policies themselves tend to amplify the phenomenon, either by lack of global information or by not using properly the global information available. In particular, the destabilizing influence of misperception of the multiple feedback loops in the case of local base-stock policies has been stressed in [3].

In terms of control system theory, load and inventory variables can be viewed as the outputs of the supply chain system, the final demands being the external inputs subject to disturbances, and replenishment policies producing the controlled inputs. From this approach, several authors [4] have studied the stability property of the system under classical base-stock policies and they have proposed in particular to reduce the nervousness of this policy by using a proportional controller.

The boundary of the control action set for a supply chain system is determined by its distributed structure. A partitioning of the causes of the bullwhip effect into structural determinants and external triggers has been proposed in [2]. More generically, this paper distinguishes internal and external factors and evaluates the possible improvements that can be expected from using relevant external data as parameters in local ordering and loading policies.

In spite of their willingness to cooperate, the partners of a supply chain generally have different constraints, objectives and practices. Local stock levels and production loads are private data that can only be integrated in local ordering policies. However, with the progress of integrated information platforms, other data can be shared by all the partners of a supply chain. In this study, it is assumed that demands for the final products of the supply chain are communicated in real time to all the partners. This assumption is technically realistic with modern communication networks and can be obtained by agreement between the supply chain partners. Then, three types of loading and ordering policies are studied: inventory-based local policies, internal-orders-based policies and external-orders-based policies. To clearly identify the influence of these policies on the performance, this study is conducted by analytical simulation with SCILAB (http://www.scilab.org) [5] on a five-product, three-level manufacturing network presented and modelled in Sect. 3.2. More generic simulation tools such as Arena (http://www.ArenaSimulation.com) could then be used to better represent specific manufacturing systems and analyse the influence of complicating factors such as varying lead times. The different ordering policies are described in Sect. 3.3. Then, Sect. 3.4 compares the simulation results, Sect. 3.5 provides some practical insights in supply chain dynamics, and some conclusions are finally presented.

3.2 A Model of a Supply Chain System

3.2.1 Presentation of the Dynamic Model

The model proposed to describe a supply chain is similar to the dynamic system presented in [6]. It is based on the product structure of the final products with the bill of materials and lead times associated with manufacturing stages. Transportation times are assumed to be included in lead times.

Classically (see, e.g. [7] and the references therein), a multi-stage production structure can be represented by an acyclic directed graph with nodes representing production activities and arcs linking components to products. The total number of product types is n. Each production activity has several input products and one output product. In the considered multi-stage production structures, there is a one-to-one correspondence between products and activities. In other words, each assembly component i ($i = 1, ..., n$) is a product that is either produced or ordered through activity i ($i = 1, ..., n$). Production of one unit of product of type i ($i = 1, ..., n$) requires the assembly of components $j = 1, ..., n$ in quantities P_{ji}. Products can be partitioned into levels. Level 0 is for end products, level l for products which are components of products of levels strictly less than l, and of at least of one level $l - 1$ product type. If products are numbered in the increasing order of their level, matrix $P = (P_{ij})$ has a lower triangular structure with zeros on the diagonal.

Each enterprise of the supply chain is supposed to perform one or several production activities using components provided by other enterprises upward in the chain and/or providing components to other enterprises downward along the chain.

As in [6], the planning horizon [0,T] is divided into time periods. The model is based on the classical assumptions of MRP (materials requirements planning) systems: product lead times are supposed constant and multiple of the elementary time-period. This assumption can be considered appropriate for planning and prediction, real lead times being estimated as their mean values obtained from statistical data.

The lead time associated with product i is denoted θ_i. In the 'gozinto' graph, constant durations θ_i are thus associated with nodes $i \in (1, ..., n)$ and arcs (i, j) are valued by P_{ij} to represent the 'gozinto' coefficients of the products.

External demands for final products are assumed to be random with constant mean values and uniformly distributed bounded variations.

The quantities involved in the model are described below for $i \in (1,..., n)$, $k \in (0,..., T)$:

- s_{ik} is the quantity of product i in stock at the beginning of period k
- v_{ik} is the quantity of product i whose production is started at period k
- z_{ik} is the quantity of product i delivered at period k
- d_{ik} is the demand for product i at period k

The demand for final products is external to the supply chain system. For primary and intermediate products, demand is the result of the ordering policy.

The system admits a dynamic representation in the form of a discrete-time multi-variable linear model with distributed lead times (considered as delays in the mathematical sense):

$$s_{i,k+1} = s_{ik} + v_{i,k-\theta_i} - z_{ik} .$$ (3.1)

Customers and producers items being delivered on stock as much and as soon as possible, delivery variables are given by:

$$z_{ik} = \min(d_{ik}, s_{ik} + v_{i,k-o_i}).$$ (3.2)

Production levels, inventory levels and backorders are subject to pointwise-in-time constraints: positivity constraints and capacity constraints. In particular, if c_{ik} is the production capacity for production of i at period k, the production orders should satisfy:

$$\sum_{l=0}^{\theta_i-1} v_{i,k-l} \leq c_{ik} .$$ (3.3)

Also, production levels may be restricted by the amounts of components available. Requirements for primary and intermediate products should be consistent with the input (or 'gozinto') matrix, P, and with production levels. These conditions are expressed by the non-negativity of inventory levels for components:

$$s_{i,k+1} \geq 0 \quad \Leftrightarrow \quad \sum_{j=1}^{n} P_{ij} v_{jk} \leq s_{ik} + v_{i,k-\theta_i}$$ (3.4)

if i is not an end-product.

3.2.2 Some Properties of the Model

The product structures considered are characterised by an output matrix which is the identity matrix (activity loads are measured in units of output products) and an input matrix P which is square, with non-negative coefficients, lower-triangular with zeros on the main diagonal. Consider the vector of mean external demands for all the products $d=[d_1 \ldots d_n]^T$. This vector being assumed constant, stability requires that the mean production vector $v=[v_1 \ldots v_n]$ is constant in steady state conditions and satisfies $d=(I-P)v$.

Then, it is not difficult to show that the inverse of matrix $(I-P)$ exists and is non-negative. It uniquely determines the mean equilibrium production vector:

$$v=(I-P)^{-1}d. \tag{3.5}$$

A vector of equilibrium inventory levels $s=[s_1 \ldots s_n]^T$ can also be defined. Its computation depends not only on the probability distributions of external demands but also on the ordering policy used for each activity. As it will be explained in the sequel, inventory positions will be used rather than inventory levels to construct efficient ordering policies.

A necessary condition for the system stability is that the equilibrium state defined by the couple (v, s) satisfies all the capacity constraints.

3.2.3 A Five-Product Example

An example of a product structure for two final products (numbered 1 and 2) is shown in Fig. 3.1.

This is a three-level product structure with final products 1 and 2 at level 0, intermediate product 3 at level 1 and primary products 4 and 5 at level 2.

It is assumed that the five products are manufactured by five different enterprises. Production capacities at each period are supposed constant and given by:

$$C=[75 \quad 60 \quad 250 \quad 740 \quad 380]^T.$$

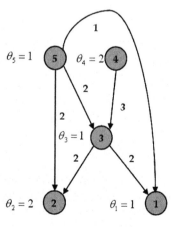

Fig. 3.1 The 'gozinto' graph of a product structure

The vector of external demands is stationary with mean value $d=[20\ 15\ 0\ 0\ 0]^T$. The corresponding bill of material, given by Eq. 3.5, gives the mean production vector:

$$v=[20\ 15\ 70\ 210\ \ \ 190]^T.$$

At any period k, d_{1k} is a uniformly distributed random variables in the interval [0, 40] and d_{2k} is a uniformly distributed random variable in the interval [0, 30]. Accordingly, equilibrium inventory levels have been computed and used to initialise the inventory trajectories when simulating the production network under the different policies.

3.3   The Production and Ordering Policies

An ordering policy is a rule that determines the amounts of components ordered to suppliers as a function of the available knowledge on the current state of the system. In a similar manner, a production policy is a rule that determines the amounts of products to be manufactured as a function of the available knowledge on the current state of the system.

Classically, it can be assumed that in a supply chain, each enterprise only knows its current and past manufacturing and supply orders, and its current and past demands (from partners and from customers). Additionally, some anticipated knowledge of future orders for each enterprise will be assumed to construct the 'order-based policy'. In this study, it is also assumed that current and past values of external demands for final products are available for all the partners of the supply chain. Such an assumption opens new possibilities for building production and ordering policies.

The three types of policies that will be compared are the inventory-based policy, the order-based policy and the demand-based policy.

3.3.1   The Inventory-Based Policy

This policy acts on the base-stock principle for the inventory position. The inventory position of product i at period k is denoted p_{ik} and is classically defined as the sum of the inventory level and pending orders:

$$p_{ik} = s_{ik} + \sum_{k=1}^{\theta_i-1} v_{t-k}.$$

The fact of using inventory positions rather than inventory levels has a smoothing effect on the ordering sequence. It clearly avoids ordering several times or not ordering for the same delivery.

A reference stock position Q_i is computed at each production stage i, and the objective of the policy is to maintain the inventory position at this reference value and at each period. So, assuming that the initial stock position is Q_i, the policy simply consists of ordering the same quantity of product i as the one requested at period k. In other words, taking into account the constraint on production capacity, each delivery triggers one production order for product i:

$$u_{ik} = \min (d_{ik}, C_i - \sum_{l=1}^{\theta_i - 1} v_{i,k-l}).$$

Component availability conditions (3.4) may limit the production level for stage i. Such conditions cannot be easily decoupled. If all the available components j can be used in priority to produce product i, then the following condition applies:

$$v_{ik} = \min_{j; P_{ji} \neq 0} (u_{ik}, \frac{s_{jk} + v_{j,k-\theta_j}}{P_{ji}}), \qquad (3.6)$$

and stock levels s_{jk} are updated by $s_{jk} = s_{jk} - P_{ji} v_{ik}$.

What is typical of this policy is that internal orders for primary and intermediate products j; $P_{ji} \neq 0$ are created from the current demands for products i through the following relation:

$$d_{jk} = \sum_{i=1}^{n} P_{ji} d_{ik}. \qquad (3.7)$$

Under this policy, demands propagate upwards without any delay, through the mechanism described by expression (3.7).

Such a policy relies on current inventory levels since they limit delivery and production at each node of the production network through condition (3.2).

3.3.2 The Order-Based Policy

The order-based policy obeys a mechanism that combines 'lot for lot' MRP and capacity constraints. Each enterprise provides products from its stock as long as the inventory level remains positive.

Backorders are avoided by anticipating production for orders that cannot be satisfied from the stock. The ordering mechanism propagates upward taking into account the bill of materials and lead times.

3.3.3 The Mean-Demand-Driven Policy

The mean-demand-driven policy is a very simple local policy in which production orders are computed from the bill of materials (3.5) and from the estimation of mean demands for final products:

$$v_k = (I - P)^{-1} \hat{d}_k .$$ (3.8)

Backorders may be generated when production capacity and/or availability of input components are not sufficient to execute these orders.

Recursive estimators of mean demands are constructed by taking the average of real demand data on a given time window denoted W:

$$\hat{d}_{ik} = \frac{W-1}{W} \hat{d}_{ik-1} + \frac{1}{W} d_{ik} .$$ (3.9)

It can be noted that due to recursive demand estimation, the mean-demand-driven policy is adaptive and can be applied to non-stationary demand profiles.

3.4   Numerical Results

In order to evaluate the bullwhip effects associated with the three policies, production and inventory variances for the five products have been estimated using series of 300 runs over a 40-period time horizon. To each run corresponds a random generation of uniformly distributed demand sequences for products 1 and 2, respectively in the intervals [0, 40] and [0, 30]. Then, model-based simulations have been performed using SCILAB [5]. The results are presented in Table 3.1. Also, for comparison purposes, the same random demand trajectories of (d_{1k}) and (d_{2k}) have been used for the three policies. These demand trajectories are displayed in Fig. 3.2. Figures 3.3–3.5 show the production and inventory evolutions for the five products under the three policies for data of Fig. 3.2.

3.4.1   The Inventory-Based Policy

A first remark concerning this policy is that it does not stabilise the system if initial inventory levels are not sufficient. In that case, backorders tend to accumulate while production levels remain insufficient due to the fact that they are conditioned by inventory levels through Eq. 3.6.

Important values of inventory levels have thus been required to obtain convergence for all the experiments. The selected initial inventory values are given by the following vector of base-stock positions: $Q = [50 \quad 55 \quad 180 \quad 750 \quad 550]$.

From simulation results of Fig. 3.3, it can be noted that the inventory-based policy sustains important variations of product load and product inventory at all the production stages.

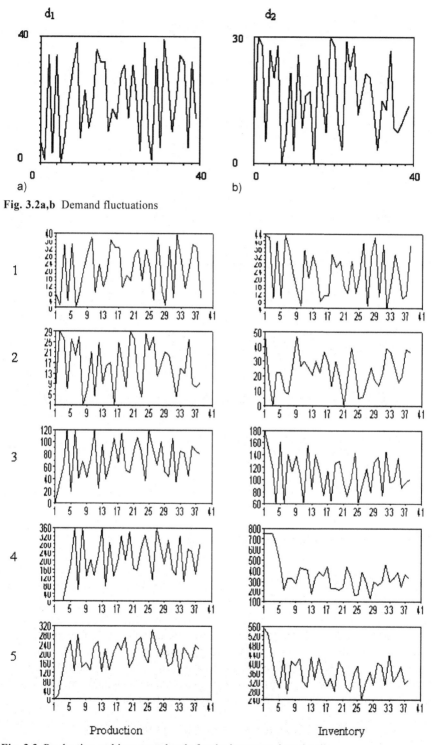

Fig. 3.2a,b Demand fluctuations

Fig. 3.3 Production and inventory levels for the inventory-based policy

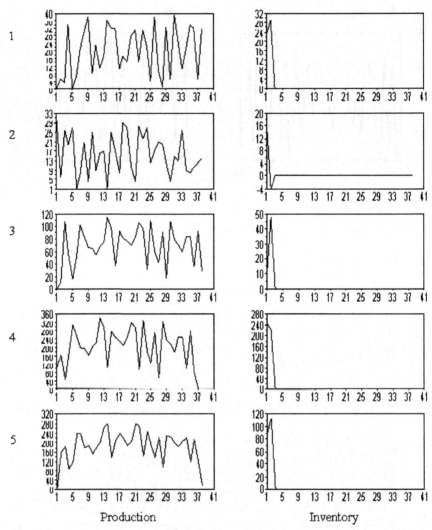

Fig. 3.4 Production and inventory levels for the order-based policy

3.4.2 The Order-Based Policy

As for the 'lot for lot' MRP technique, the order-based policy uses the initial stock in the transient stage and then, stocks are maintained at the zero level, production anticipating demand. Figure 3.4 shows that better stability results have been obtained for this policy in terms of demand satisfaction. Compared to the inventory-based policy, lower initial inventory levels have been needed:

$$S=[30 \quad 25 \quad 70 \quad 250 \quad 150].$$

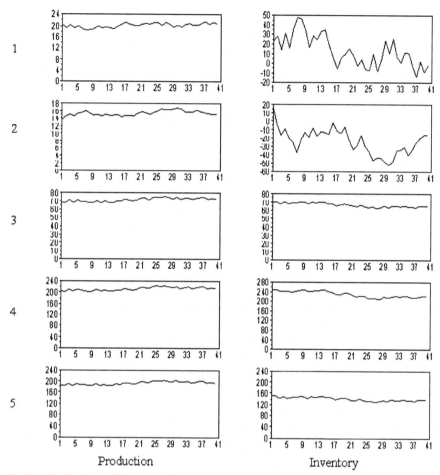

Fig. 3.5 Production and inventory levels for the mean-demand-driven policy

With this policy, inventory levels are completely stabilised in a few periods but production fluctuations remain important, of the same order of magnitude as for the inventory-based policy.

3.4.3   The Mean-Demand-Driven Policy

Results of Fig. 3.5 show that under the mean-demand-driven policy, demand fluctuations are almost completely absorbed by the inventories of final products. Smooth production curves are obtained for all the enterprises. Inventory variations are very limited for primary and intermediate products. Important inventories are needed for final products but they can be kept very small for primary and intermediate

Table 3.1 Variance comparison for the three policies

	Products	1	2	3	4	5
	Demand variance	134	75	0	0	0
Inventory-based policy	Production variance	11	20	100	1500	600
	Inventory variance	18	10	60	2500	600
Order-based policy	Production variance	135	75	850	10000	4500
	Inventory variance	1	1	3	100	30
Mean demand-driven policy	Production variance	1	1	8	80	60
	Inventory variance	500	300	10	300	60

products. Note that the amounts of backorders for final products are represented as negative inventories.

3.4.4 Variance Analysis for the Three Policies

The three policies have been simulated on series of 300 randomly generated demand trajectories for final products 1 and 2. The mean demand values have been kept constant ($d_1 = 20$, $d_2 = 15$). Demand variances associated with the uniform distributions have been evaluated and indicated in Table 3.1. These variances are compared with production and inventory variances for the five products. In interpreting the results, it is important to take into account the bill of materials, which naturally amplifies for components the means and variances of production and inventory levels.

From this table, it is clear that under the inventory-based policy, demand variations generate important production and inventory fluctuations along the supply chain. The order-based policy damps oscillations of inventory levels but not those of production levels. The mean-demand-based policy considerably attenuates the final demand disturbances at all the stages of the supply chain, except for the final stage, for which important inventories are needed.

3.5 Supply Chain Dynamics in Practice

3.5.1 The Beer Game

Playing the beer game is one of the best exercises for understanding the supply chain dynamics generated by fluctuating demand and/or disrupted supply. Several generations of students and managers have played this game to perceive the risk of amplifying fluctuations by applying local and/or shortsighted ordering policies. This game also provides a good benchmark for comparing the ordering policies described in this chapter. In this game, the players act as the different enterprises of a supply chain that produces, distributes and sells beer. The aim of each player is to minimise his total cost (inventory cost+backlog cost) over a time horizon of, say, one year, divided into periods of, say, one week. At the beginning of each period, the retailer receives customer orders, which are randomly generated. He then places an order to the wholesaler, who places an order to the distributor, who, in turn, places an order to the manufacturer. Ordered products are then delivered, stage by stage, with lead times that may be constant or random. The game is usually played in a distributed manner, each player acting as an independent enterprise with its own ordering policy. To play this game, several simulators are available on the internet (http://beergame.mit.edu, http://www.masystem.com/beergame).

3.5.2 Some Real Consequences of the Bullwhip Effect

Consequences of the bullwhip effect are generally measured in terms of costs: holding costs associated with on-hand inventory and backorder costs that penalise the delay in serving demand. However, the mathematical cost functions that are classically used to represent these costs cannot fully grasp the non-linear and discontinuous nature of real costs. If existing storage space is insufficient to cover the risk of shortage, new investments are required, causing a discontinuity in inventory costs. Conversely, shortage cost functions are generally unable to capture the risk of losing a market, and possibly the critical risk of bankruptcy for the weakest links in the chain. Such risks are particularly important for SMEs (small and medium enterprises) involved in supply chains. As stated in [8], joining a supply chain is more risky for an SME than for a large company since it has to invest most of its assets and often has to use investment loans to catch up with the high quality and efficiency standards of the supply chain. A European coordinating action, named CODESNET (Collaborative Demand and Supply Networks) has constructed a virtual laboratory and a virtual library to describe and analyse enterprise networks, and particularly networks of SMEs (http://codesnet.polito.it). The laboratory provides many real examples of enterprise networks in Europe and describes their operation structure, interaction with socio-economic environment and organisation arrangement. One of the potential uses of this web portal is to help SMEs to limit their risks and costs through collaborative networking, including coordination and control of their product flow dynamics.

3.6 Conclusion

Fluctuations of production and inventory levels are typical of multi-level material replenishment processes with significant lead times and fluctuating demands. In supply chains, this phenomenon, known as the bullwhip effect, has been observed in many practical cases. By simulation of analytical models, the study has shown that a combination of classical local ordering policies tends to sustain or even amplify this effect. Two classical local policies have been studied: an inventory-based policy, of the base-stock type, and an order-based policy, inspired by MRP. In both cases, a combination of local feedback loops reinforces the oscillations and may generate instability. Conversely, smoother production and inventory levels have been observed under the so-called mean-demand-driven policy. Under this policy, the partner enterprises do not directly influence each other. They rather respond to integrated (or averaged) demand evolutions in parallel, global consistency in production and assembly being achieved through respect of the bills of materials.

3.7 Questions and Assignments

- What is the bullwhip effect?
- What are the main causes of the bullwhip effect?
- What is the game often played to illustrate the bullwhip effect?
- What are the main levers to decrease load and inventory fluctuations in supply chains?
- In the numerical examples, high initial inventory levels have been assumed. Suppose now that initial stocks are empty for all the products. What is the minimal ramp-up time, which is the minimal number of periods necessary to obtain the reference initial inventory levels, denoted Q_1, Q_2, Q_3, Q_4, Q_5? Assuming that the system is over-capacitated, give a production schedule to obtain these reference levels.

References

[1] Lee LH, Padmanabhan P, Whang S (1997) Information distortion in a supply chain: the bullwhip effect. Manag Sci 43(4):546–558
[2] Miragliotta G (2004) The bullwhip effect: a survey on available knowledge and a new taxonomy of inherent determinants and external triggers. Preprints 13th WSPE 3:269–273
[3] Sterman J (1989) Modeling managerial behavior: misperceptions of feedback in a dynamic decision making experiment. Manag Sci 35(3):321–339
[4] Dejonckheere J, Disney SM, Lambrecht MR et al (2003) Measuring and avoiding the bullwhip effect: a control theoretic approach. Eur J Oper Res 147(3):567–590
[5] Gomez C (1999) Engineering and scientific computing with Scilab. Birkhaüser, Basel
[6] Hennet J-C (2003) A bimodal scheme for multi-stage production and inventory control. Automatica 39:793–805
[7] Muckstadt GA, Roundy RO (1993) Analysis of multistage production systems. In: Graves SC, Rinnooy Kan AHG, Zipkin PH (eds) Handbooks in operations research and management science, vol 4. North-Holland, Amsterdam
[8] Hennet J-C, Mercantini J-M, Demongodin I (2008) Toward an integration of risk analysis in supply chain assessment. In: Proceedings I3M-EMSS'08:255–260

Chapter 4
Pharmaceutical Distribution Network

H. Van Landeghem

Abstract This chapter describes the tactical organisation of a distribution network of pharmaceutical products. It focuses on the optimal selection of transportation modes, i.e. the size and frequency of shipments, from a central storage location to all markets inside Europe. Tactical organisation means that the objective is to determine the characteristics of the shipments, to serve as a basis for contract negotiations with the intended carriers. The problem is modelled as a mixed-integer linear program, and solved accordingly. Then simulation is used to check the robustness of the optimal results. The results indicate a potential reduction of total transportation costs of 36%. Simulation indicates a deviation of only 4%. Its main characteristics are a reduced shipping frequency for small customers, indicating that marketing will have to be involved in the practical implementation. Additional strategic options, regarding adding distribution centres (DCs), were also investigated with the same model.

4.1 The Pharmaceutical Distribution Network

The research was performed for a large company, producing pharmaceutical consumables and measurement instruments, as part of a Master's thesis [1]. The focus was on the European distribution network, representing some 30% of total revenue. Customers were served from one central European distribution centre (EDC), located in Belgium. This EDC received inbound goods from 25 different plants located across the world. Outbound shipments to customers over the road from this EDC carried some 79 different product types (representing 25,000 SKUs) to many

Hendrik Van Landeghem
Ghent University, Belgium
hendrik.vanlandeghem@ugent.be

countries in Europe (see Table 4.1). Total transportation volume amounts to about 377,900 m³ per year, for a transportation cost of roughly 21 million euros. The contribution of the main product types to this volume is shown in Fig. 4.1. Product type names have been removed for confidentiality reasons. It clearly reveals the complexity of determining cost-optimal transportation modes: some low-volume products have higher costs, because of small shipment sizes at high unit costs (package or small pallet tariffs, as used further on).

The first step in research is always to transform the available information into a meaningful and manageable data set. Historical shipment information of a typical year was available in electronic format. We aggregated the thousands of business customers (hospitals, doctors' surgeries, importers) into 739 geographically coherent 'zones' (see Table 4.1) in order to obtain a finely meshed demand model that would faithfully represent both transportation destinations and distances. This was needed because the optimisation model had to be allowed to group different product flows into combined shipments, and then select the optimal transportation mode for this flow. Figure 4.2 shows the location of major customers (representing 80% of total demand).

The research questions we cover in this chapter are:
1. What is the optimal mode of transportation for inbound and outbound flows?
2. How will this evolve in a growing business for the next 5 years?
3. What will be the effect of adding a second echelon of local DCs, specifically for the German regions?

Table 4.1 Number of customer zones per country in the study

Country	No. of zones	Country	No. of zones
Austria	11	Greece	1
Belgium	9	Ireland	2
Germany	86	Italy	92
Denmark	9	Luxemburg	1
Spain	52	Netherlands	10
Finland	99	Norway	10
France	97	Portugal	3
Great Britain	174	Sweden	83
Total no. of zones			**739**

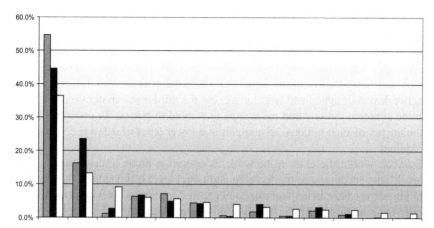

Fig. 4.1 Product mix (volume, weight, transportation cost respectively)

Fig. 4.2 Pharmaceutics distribution network with major clients

4.2   Determining Transportation Modes
and Associated Unit Costs

Logistical networks can be designed and analysed on three different levels [2]:

- *Strategic*: determines the structure of the network (echelons, number of sites, type and functionality of the logistical platforms at each site). In this case the network was designed and functioning, but the third research question related to some strategic options that the company wanted to pursue. In order to analyse the impact of such options, an analysis at a lower lever of detail (i.e. tactical) is required.
- *Tactical*: within the existing network its operating mode is determined at this level. This encompasses the location and target level of inventory, the transportation modes and carrier fleet mix, and the SLAs (service level agreements) that one wants to pursue. The research in this chapter was carried out at the tactical level.
- *Operational*: the lowest level of analysis focuses on the day-to-day execution of the logistic tasks. Analytical tools such as schedulers and routing optimisation try to minimise operating costs within the framework of the tactical and strategic decisions already made.

The system boundaries that were studied in this research consisted of the transportation modes used in the inbound and outbound goods flow. The time frame was 5 years. Since we wanted to know the transportation mix, i.e. the mix of carrier types and capacities, that the company needed to contract for the following years, we chose one month as unit of time for analysis. This time unit was a trade-off to guarantee enough variation over the 5 years of study subject to the changing demand patterns, but at the same time avoid getting into a cumbersome and detailed analysis on a weekly or daily level. This would probably lead to intractable models anyway.

One of the most difficult aspects of any study is data collection. In logistic models the unit costs for transportation are very important, since optimal results are very sensitive to variations in these costs. In practice, however, the cost of transporting 1 m³ between any two locations is very hard to obtain. The carrier landscape in Europe is of a staggering complexity, with an abundance of potential suppliers, and a bewildering array of different fee structures for transportation. Obtaining transportation costs for this study was split into two parts: unit costs for the current transportation links, and costs for links that would appear in the above scenarios, but were not used today (hence no actual costs were available inside the company).

We first had to determine the different transportation modes that were in use within the company. A transportation mode is one form of carrier, with an associated maximum volume and a cost per volume per kilometre. An 'FTL_Road' Mode (FTL = full truck load) e.g. has a maximum volume of 70 m³.

4.2.1 Inbound Transportation

The inbound flow to the EDC originated from a total of 25 plants (or aggregated suppliers). Inbound flows were transported by sea (container), air and road. Figure 4.3 shows the volumes per shipment that were shipped by sea mode, from one UK plant to the EDC. It has a bimodal distribution: most shipments consist of either one pallet of 5 m³ or of a near-truckload of 50 to 70 m³. Analysis of the size of shipments revealed a global utilisation rate of 77% for FTL truck shipments, indicating a first possibility to reduce costs by increasing utilisation. Since FTL rates are regardless of the truck utilisation, increasing this factor will immediately reduce the transportation cost. It will, however, have an impact on the frequency of shipments and the delay. Both cannot be determined without deeper analysis, which was the first purpose of this study.

Rates for air transportation are even more complicated. Besides being much higher per unit of weight, volumetric shipments (such as those in this company) are sometimes penalised. One carrier, e.g. assigns a minimal weight of 166 kg to 1 m³ to determine the transportation cost. Air transportation was used to make up for delays in delivery due to previous shortages in the supply chain.

4.2.2 Outbound Transportation

Due to the sheer number of outbound shipments, we first analysed the data in bulk. We discovered that trucks contracted at FTL rates were not always fully loaded, leading to higher unit transportation costs for the goods involved. Figure 4.4 shows

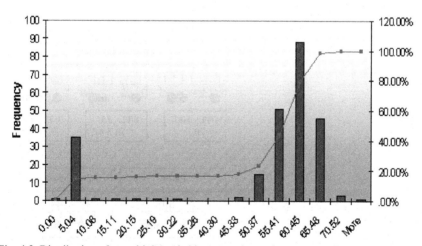

Fig. 4.3 Distribution of actual inbound shipment volumes in sea mode from UK plant

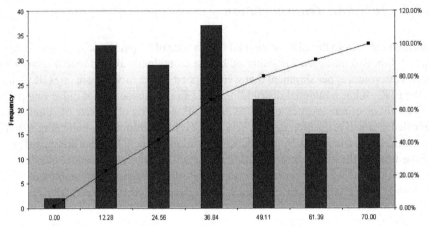

Fig. 4.4 Actual shipment volumes at FTL cost

the distribution of outbound shipment volumes using FTL carriers. It reveals an average truck load of 36.2 m³ or 52% utilisation. Cost figures were fairly easy to obtain from the company's books. The FTL cost of this (existing) link was about 1,050 €, regardless of utilisation.

To keep the model tractable, we decided to map all shipments into a limited set of load modes. A *load mode* is a certain level of loading for each transportation mode, together representing a simplified set of transportation choices. The load modes for truck transport with their volume capacity are listed in Fig. 4.5. The total cost of an FTL was thus allocated to the actual load, yielding a truck load mode 'FTL_100' with unit cost of 15 €/m³ when loaded 100%, increasing to 20 €/m³ for FTL_75 (loaded at 75% capacity) and 30 €/m³ for FTL_50.

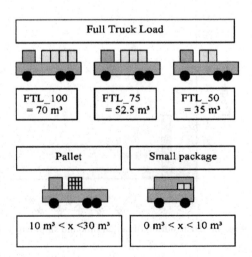

Fig. 4.5 Loading modes for road-bound transportation

Using a similar approach we also defined lane modes. A *lane mode* is a load mode, carried out with a specific frequency within the monthly time frame of the analysis. A lane mode of 'FTL_10_75' indicates a truck running 10 times a month (on average every 3 days) loaded at 75% of capacity at FTL rate. In this way we obtained a total of 22 outbound lane modes (Table 4.2), and 3 inbound ones, based on the 5 load modes of Fig. 4.5. This approach is an extension of [3], which uses only 2 truck types.

To determine the correct cost per kilometre for the different modes on new links, we investigated the costs of multiple carriers (both current and prospective ones) using linear regression. This obtained a fairly reliable result as shown for the 'pallet' lane mode in Fig. 4.6.

Table 4.2 Outbound lane modes used in the MILP model

Lane mode	Frequency	Volume shipped
FTL_1_100	1 × /month	70 m³
FTL_1_75	1 × /month	52.5 m³
FTL_1_50	1 × /month	35 m³
FTL_2_100	2 × /month	140 m³
FTL_2_75	2 × /month	105 m³
FTL_2_50	2 × /month	70 m³
FTL_4_100	4 × /month	280 m³
FTL_4_75	4 × /month	210 m³
FTL_4_50	4 × /month	140 m³
FTL_10_100	10 × /month	700 m³
FTL_10_75	10 × /month	525 m³
FTL_10_50	10 × /month	350 m³
FTL_15_100	15 × /month	1050 m³
FTL_15_75	15 × /month	787.5 m³
FTL_15_50	15 × /month	525 m³
FTL_30_100	30 × /month	2100 m³
FTL_30_75	30 × /month	1575 m³
FTL_30_50	30 × /month	1050 m³
Drop_FTL_100	n.a.	70 m³
Drop_SP	n.a.	0 to 30 m³
Pallet	n.a.	10 to 30 m³
Small_Package	n.a.	0 to 10 m³

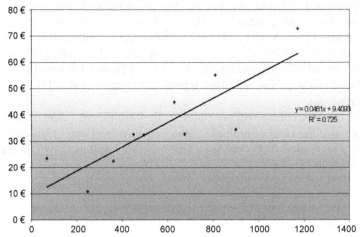

Fig. 4.6 Regression to determine cost/km of a pallet on new transportation links

4.3 Lane Mode Optimisation Model Using MILP

We selected an MILP (mixed integer linear programming) model to optimise the transportation allocations. The underlying model is a *multi-commodity flow problem*, with duplicated flow links between origin and destination nodes for each combination of product and lane mode. This model is used regularly for real-world problems, as shown recently in [4]. The flow costs are associated with the lane mode arcs as explained before and capacity bounds ensure that the total flow over all products for each link fall inside the correct lane mode. For a realistic problem such as this one, the network becomes extremely large, and solution times will be impressive unless one uses specialised algorithms that exploit the structure of the problem [5–7]. However, due to time constraints we opted for the standard branch-and-bound approach.

Figure 4.7 shows the distribution network schematically. The model will allocate flow to cover the product demand per zone (Pr A–E in Fig. 4.7) and select the optimal lane modes for it. This model is of the so-called invariant type: it contains no time expansion and therefore calculates the steady-state behaviour during a given reference period. We chose this reference period to be one month, as the total amount during this period had to be sensitive enough to the transport mode, yet be large enough to encompass all products, also the low-volume ones. A reference period of one week would have resulted in volumes that are too low for many products, leading to the erroneous choice of pallet modes. Conversely, a reference period of a year would result in large amounts of any product, but no longer with any correlation with their transport mode (because we cannot impose a priori the number of trips during a year).

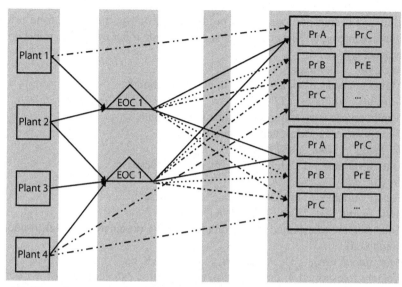

Fig. 4.7 Network representation of the shipment allocation model

4.3.1 MILP Model Formulation

We now formulate the optimisation model.

First we define the *sets of variables*:

1. $PL = \{pl : pl = 1 \dots \# plants\}$ Set of sourcing plants (25)
2. $EDC = \{edc : edc = 1 \dots \#edc\}$ European distribution centres (currently only 1)
3. $C = \{c : c = 1 \dots \#c\}$ Customer zones (739)
4. $PR = \{pr : pr = 1 \dots \#pr\}$ Product families (79)
5. $SITES = \{si : si = 1 \dots \#si\}$ Nodes in our distribution network
6. $TT = \{tt : tt = 1 \dots \#tt\}$ Lane modes (25 outbound, 3 inbound)
7. $LINKS = \{link_{o,d,tt} : link_{o,d,tt} \in \{SITES \cup SITES \cup TT\}\}$
   Set of network links for each origin–destination–lane mode combination
8. $LINKS\_ext = \{linkext_{o,d,tt,p} : linkext_{o,d,tt,p} \in \{SITES \cup SITES \cup TT \cup PR\}\}$
   Set of network links for each origin–destination–lane mode–product combination
9. $EDC\_PR = \{(edc, pr) : edc \in EDC, pr \in PR\}$
   Distribution nodes, subset of SITES
10. $C\_PR = \{(c, pr) : c \in C, pr \in PR\}$
    Set of product–customer zone combination

The following *parameters* (constants) are used in the model:

11. $Capacity\_EDC_{edc} \geq 0 : edc \in EDC$
    The capacity in m³ of each distribution centre
12. $D_{c,pr} \geq 0 : c \in C; pr \in PR$
    Total demand of customer zone c for product pr in a typical month

13. $TC_{link_{o,d,tt}} \geq 0 : link_{o,d,tt} \in LINKS$

Unit cost to transport one unit of product using transport mode tt, from origin o to destination d. This cost depends on the mode tt but not on the transported product type pr.

14. $MaxCap_{lane} \geq 0 : lane \in TT$

Maximum capacity of a transport mode in one month

15. $MinCap_{lane} \geq 0 : lane \in TT$

Minimum capacity of transport mode in one month

The *variables* in the model are:

16. $In\_use_{link_{o,d,tt}} : link_{o,d,tt} \in LINKS$

Binary variable, $=1$ if link (o,d,tt) is used, $=0$ otherwise

17. $In\_use\_edc_{edc} : edc \in EDC$

Binary variable, $=1$ if EDC is open, $=0$ otherwise

18. $Mov\_quant_{link_{si,si2,tt}} : link_{si,si2,tt} \in LINKS$

Total volume transported in one month over the designated link (all products combined)

19. $Mov\_quant\_ext_{link\_ext_{o,d,tt,pr}} : link\_ext_{o,d,tt,pr} \in LINKS\_ext$

Volume transported in one month over designated extended link of a specific product, the real decision variables

The model can now be formulated mathematically.

The *objective function* minimises the total cost of all flows through the active lane mode links. The EDCs are not open/closed by the optimisation model, given the uncertainty over the opening costs. Instead, we will set the variables *In_use_edc* according to the scenario investigated. The objective function is formulated as follows:

20. $\forall si \in SI, si2 \in SI, pr \in PR, tt \in TT, edc \in EDC : links\_ext_{si,si2,tt,pp}$

$$Min(TotalCost) = \sum_{links_{si,si2,tt}} Mov\_quant_{links_{si,si2,tt}} * TC_{links_{si,si2,tt}}$$

The objective function is optimised subject to the following *constraints*:

21. $\forall edc \in EDC :$

$$\sum_{link\_ext_{edc,si,tt,p}} Mov\_quant\_ext_{link\_ext_{edc,si,tt,p}} \leq 10\,000\,000 * In\_use\_EDC_{edc}$$

22. $\forall (edc, pr) \in EDC\_PR :$

$$\sum_{links\_ext_{o,edc,tt,p}} Mov\_quant\_ext_{link\_ext_{o,edc,tt,p}} = \sum_{links\_ext_{edc,d,tt,p}} Mov\_quant\_ext_{link\_ext_{edc,d,tt,p}}$$

23. $\forall link_{(si,si2,tt)} \in LINKS :$

$$Mov\_quant_{link_{si,si2,tt}} = \sum_{links\_ext_{si,si2,tt,p}} Mov\_quant\_ext_{links\_ext_{si,si2,tt,p}}$$

24. $\forall (c, pr) \in C\_PR:$
$$\sum_{link\_ext_{si,c,tt,p}} Mov\_quant\_ext_{link\_ext_{si,c,tt,p}} = D_{c,pr}$$

25. $\forall link_{(si,si2,tt)} \in LINK : \sum_{link_{si,si2,tt}} Mov\_quant_{link_{si,si2,tt}} \geq MinCap_{lane} * In\_use_{link_{si,si2,tt}}$

$\forall link_{(si,si2,tt)} \in LINK : \sum_{link_{si,si2,tt}} Mov\_quant_{link_{si,si2,tt}} \leq MaxCap_{lane} * In\_use_{link_{si,si2,tt}}$

The constraints are explained below:

21. This only allows flow through an EDC if it is open. The large factor we used can be replaced by the total throughput capacity of the facility, but this was not used in our scenarios.
22. This is the flow conservation constraint, forcing all products arriving in an EDC node to leave. Note that our network has thus only three echelons: suppliers, EDCs and customer zones.
23. These constraints are so-called 'bundle constraints': they combine all volumes of each product, transported over a certain link, into the same transport mode. We enforce equality to make sure the flow quantity through the lane mode selected matches both its lower and upper capacity bounds.
24. The constraints that force total flow per product to equate the customer demand in that month. Equality is used, since we do not want the model to over-deliver product to the customer in order to benefit from a cheaper, but more fully utilised transportation mode.
25. Each transport mode has a valid range of lower and upper capacity in one month. Each lane mode is thus defined by a dual set of constraints to obtain the desired allocation to the available transportation modes.

The model was programmed using AMPL and solved with CPLEX solver on a Pentium 4 computer. Due to the sheer size of the model, we had to split it into six different geographical regions (that did not interact), and we ran the branch-and-bound procedure until we reached a gap of 5% from optimality. This proved to be acceptable to the company.

4.3.2 Experimental Design and Results

We solved the model for different scenarios and data sets. First we solved the model using the current demand and current flows (fixed by adding constraints). The model turned out to be very accurate: its total cost for the current transportation situation was 101.3% of the real cost, as shown in Table 4.3. Deviations per country were larger, but well within acceptable limits, given that the model optimises a transportation mix decision (and not individual shipments). This proved that our simplification using the lane mode concept was sufficiently close to the real situation.

Table 4.3 Model validation results

Country zone	Model vs. reality
Belgium	101.7%
Denmark	99.4%
Germany	88.9%
Finland	99.0%
France	112.0%
Great-Britain	107.1%
Greece	99.2%
Ireland	99.5%
Italy	101.6%
Luxemburg	97.3%
Netherlands	99.3%
Norway	99.1%
Austria	99.3%
Portugal	98.7%
Spain	100.0%
Sweden	99.0%
Total cost	**101.3%**

Subsequently we reverted to simulation to measure the validity of the optimal transportation mix under different demand situations. We built a simulation model in MS Excel that would use the optimal mix, and tried to fit different demand instances into it. The number of shipments was adjusted to accommodate the total demand, if needed. We generated five different instances of monthly demand for each customer zone, according to the distribution of the real demand. The standard deviation of the total transportation cost for all instances, measured by the coefficient of variation (i.e. standard deviation divided by average) was 3.9% for the total distribution network, which led us to conclude that the numerical results were sufficiently robust for the demand patterns that occurred within the company. Table 4.4 shows more details per country.

The first results indicated an immediate potential reduction by 37% of total transportation costs for the current demand pattern. This reduction was obtained through more fully loaded outbound FTL shipments, with a slightly lower frequency. To realise these savings however, the company has to convince small customers to accept a lower frequency of replenishment. In any case, marketing now has a good idea of the implied cost of the current customer service policy. Figure 4.8 shows the optimal lane mode mix for a small Portuguese customer.

In year 1 it consists primarily of pallets and a monthly FTL shipment that is 50% loaded. As his volume increases over the next 5 years, the (optimal) transporta-

Table 4.4 Robustness of solution validated by simulation

	Coefficient of variation
Germany	3.4%
Italy	7.6%
Spain	4.9%
EURREST	6.4%
France	3.1%
Great Britain	4.2%
NORDIC	5.0%
Total	**3.9%**

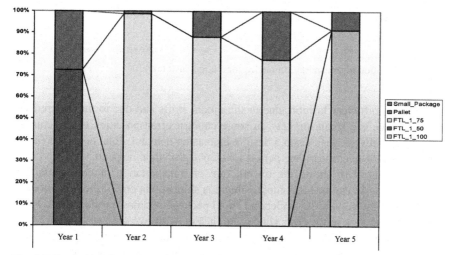

Fig. 4.8 Evolution of transportation modes for a small customer

tion mix shifts to one monthly shipment of a 75% loaded FTL, with some small packages in between. Only in the final year of the analysis will the customer have enough volume to warrant a monthly shipment of a fully loaded truck, yielding the lowest unit cost.

In Fig. 4.9 the same is shown for a large French customer, who is served by a monthly FTL shipment at 100% and 50% respectively, weekly FTL shipments at 75%, some drop shipments and a small amount of pallets. Drop shipments are direct shipments from the plant to the customer, not through the EDC. Over the 5-year demand evolution the transportation mix evolves towards weekly shipments of a 100% FTL truck and a 50% FTL, the rest being delivered by drop shipments.

Following this we analysed the total transportation costs per country. Although the overall demand was projected to grow at a fairly constant rate (though a different rate in different countries), the resulting transportation costs themselves evolve

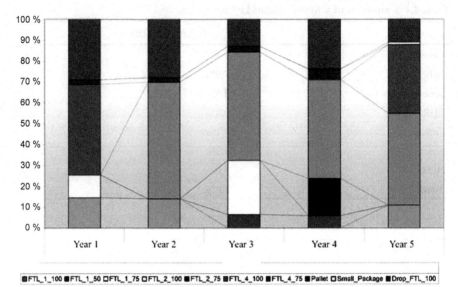

FTL_1_100 FTL_1_50 FTL_1_75 FTL_2_100 FTL_2_75 FTL_4_100 FTL_4_75 Pallet Small_Package Drop_FTL_100

Fig. 4.9 Evolution of transportation modes for a large customer

in a highly non-linear fashion, due to the modal shifts and due to the differences between large and small markets. In some countries transportation costs actually went down between year 4 and 5 despite increasing demand. Total costs increased with about 11% over the 5-year period for a 36% growth in demand.

Finally, Fig. 4.10 shows the overall mix of transportation modes across the whole market. As can be seen, most volume is shipped via one of the FTL modes. Despite this, the volume via pallet and small package remains considerable. However, when we look at the costs (line graph in Fig. 4.10) we see that the latter two generate roughly two-thirds of the total cost! This result was rather startling for the company's logistic managers. It made them think hard about their policy of shipping any quantity to their customers within a couple of days after order receipt. The model will allow them to carefully cost out any alternative delivery policies that they will discuss with marketing.

The same model (with minor adaptations in the topography) was then used to evaluate a scenario in which three local warehouses were to be installed in Germany, effectively adding an echelon between the EDC and the German customer zones. The effect of sending more FTL shipments closer to the German customers delivered a further 14% reduction in total transportation costs (on top of the 36% of previous scenario). These savings will have to be compared to the cost of opening and running the warehouses, which fell outside the scope of this research.

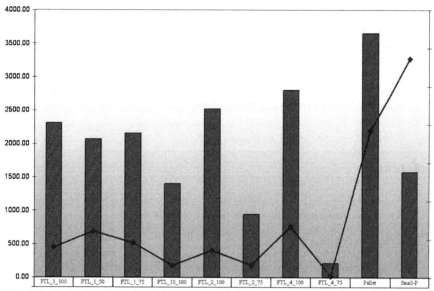

Fig. 4.10 Optimal outbound volume and cost per lane mode (*bar* volume, *line* cost)

4.4 Conclusions

This chapter describes the use of an MILP model to optimise the transportation modes for a pharmaceutical distribution network. It has demonstrated that reliable unit transportation costs can be obtained and that judiciously discretising transportation into so-called lane modes maintains validity and accuracy of the total solution. The optimal solution has the potential to lower total costs by 36%, indicating the needed shift in truck utilisation and transport frequencies. The evolution of total distribution costs under growing demand showed a highly non-linear increase over a 5-year period, illustrating the advantage of sophisticated models over simple extrapolation.

4.5 Assignments

1. For your region, find out the cost of rail and road transportation for a unit load of 1 m³ and for 1 kg. What is the unit cost for goods with high volume and low weight, and vice versa? Try to determine the underlying structure for this tariff structure.
2. The MILP model takes the total shipment volume between an origin–destination point, and tries to allocate it optimally over the available lane mode capacities. This strongly resembles the knapsack problem, a well-known operations research model. How does the knapsack problem resemble this transportation

model, and how does it differ? Find a heuristic or construct one yourself that solves this knapsack problem quickly and approximately.
3. What approach should marketing follow to convince small customers to accept lower frequency shipments per month? Are there alternatives to keeping the transportation cost low, without sacrificing frequency? What would be the main implications for the customer when the shipment frequency is reduced, e.g. from four times per month to once per month?

References

[1] De Smet J, De Vuyst K (2005) Modelling and analysing the logistic network of a pharmaceutical producer. Master's thesis (in Dutch), Ghent University, Ghent
[2] Van Landeghem H, Vanmaele H (2002) Robust planning, a new paradigm for demand chain planning. J Oper Manag 20:769–783
[3] Zäpfel G, Wasner M (2002) Planning and optimization of hub-and-spoke transportation networks of cooperative third-party logistics providers. Int J Prod Econ 78:207–220
[4] Li H, Tian S, Pan Y et al (2005) Minimum-cost optimization in multicommodity logistic chain network. Comp Algebra Geom Algebra Appl Lect Notes Comp Sci 3519:97–104
[5] Goldberg A, Oldham J, Plotkin S et al (1998) An implementation of a combinatorial ap-proximation algorithm for minimum-cost multicommodity flow. Integer Program Comb Optim Lect Notes Comp Sci 1412:338–352
[6] Holmberg K, Yuan D (2000) A Langragian heuristic based branch-and-bound approach for the capacitated network design problem. Oper Res 48(3):461–481
[7] Wu S, Golbasi H (2004) Multi-item, multi-facility supply chain planning: Models, complexities and algorithms. Comp Optim Appl 28(3):325–356

Chapter 5
Hospital Resource Management

R. M. Aguilar Chinea, I. Castilla Rodríguez and R. C. Muñoz González

Abstract In this study, a discrete-event simulation conceptual model and a set of simulation models have been developed for analysing a resource distribution problem in a hospital. The goal of the simulation has been to study the suitability of simulation as a tool which helps to better understand a real system. The models presented simulate different resource distributions. Thus, better knowledge about resource and process inter-relations in the hospital is obtained. This experience is part of a larger study on the use of simulation in the decision-making processes in hospital resource management problems.

5.1 Introduction: Objectives

A hospital, whether public or private, can be seen as a social system. As such it has evolved more slowly than other organisations, in large part due to the lack of a working definition for its product. The field of hospital management has come into its own recently, and its importance is bound to increase even more in the coming years.

One of the biggest challenges facing health care systems in industrialised nations is how to face the expectations and demands placed on them by society. The progressive development of health care systems, along with an ageing population, a higher standard of living, and scientific and technological advancements in the fields of medicine and treatment of disease, have produced a considerable increase in the resources dedicated to health care in many countries. Add to this the changes in mortality and morbidity rates, and the appearance of new chronic and crippling

Rosa M. Aguilar Chinea, Iván Castilla Rodríguez and Roberto C. Muñoz González
Universidad de La Laguna, Spain
raguilar@ull.es, ivan@isaatc.ull.es, rmglez@isaatc.ull.es

Y. Merkuryev et al. (eds.), *Simulation-Based Case Studies in Logistics*
© Springer 2009

diseases, the resolution of which requires the combined efforts of health and social services.

Within this context it may seem trite to state that health care resources, like those of other social services, are limited, and that the demand for health care is not only unlimited, but likely to grow exponentially, due more to increasing social expectations than to the emergence of new health care techniques or procedures.

With the development, both technical and structural, of health care systems, and as citizens have become better informed, the demand for health care has grown at an increasing rate, such that the demand will always outpace the ability of health care systems, and the resources assigned to them by society, to respond. In light of this situation, which results in a tension between the available resources and the expressed needs, the compromise reached by health care authorities and managers has been a moderate and orderly increase in health care spending within the constraints of the national budget, as well as an improvement in the system's productivity.

The solution to this problem in different countries has basically been to limit health care benefits in order to diminish the increases in demand, to involve health care professionals in assigning resources, and lastly, to establish efficiency standards so as to maximise the available resources. Due to the limited amount of available resources, patients create waiting queues. Waiting queues, and more specifically, the time the patients must spend in these queues before being attended, are one of the commonest used hospital's key performance indicators (KPIs). A waiting queue represents either a physical queue, where the patients physically wait to be attended, or a functional queue, such as a waiting list. To meet an acceptable KPI value, the use of simulation is proposed as a tool to obtaining a deeper knowledge of the processes that take place in hospitals, thus allowing the decisions pertinent to resource redistribution to be made with the a priori knowledge of the effect such decisions will have throughout the hospital.

In what follows, we describe a management tool which allows for the modelling and simulation of a hospital. With it, the administrator will have a more exhaustive knowledge of the system involved, the objective being to reduce the uncertainty that exists when making resource allocation decisions. The different simulations will answer the following question:

Will the efficiency of the hospital improve if a certain resource is increased/ decreased?

5.2 Conceptual Model

Since a hospital is a very complex system, we modelled it by resorting to a reduction strategy (divide and conquer), reducing the system into a group of less complex subsystems. Each identified subsystem is then modelled which, when properly reassembled into a whole, yields a global model of the hospital.

To identify the different subsystems involved in a hospital, we drew upon their functionality, resulting in a hospital made up of the three subsystems shown in Fig. 5.1.

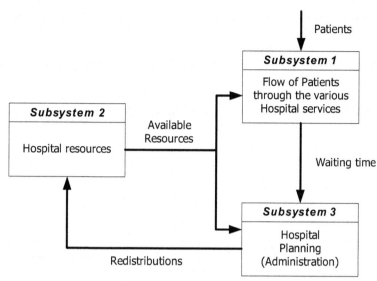

Fig. 5.1 Hospital block diagram

Subsystem 1, which represents the flow of patients in the hospital, has as an uncontrolled input (perturbation) the arrival of patients and as a controlled input the actions the administrator takes to bring the hospital to a desired state. The output is the aggregate of the variables which indicate the state of the hospital. Given that the goal of a hospital is to treat all incoming patients, the variables which indicate its proper operation (KPIs) are the average time and the mean deviations from that time that patients must wait to be attended. That is why the state of this system, from a management point of view, is defined by the average waiting time of all the queues generated.

Subsystem 2 models the resources available for attending the patients. Its input comes from the redistribution of resources made by subsystem 3 to correct inefficiencies in the hospital. Its output is the total sum of resources available at any given time. All this information is stored in a database.

Subsystem 3 describes the hospital management. Its inputs are the different queues and the resources available to maintain or improve the hospital's performance. Its output produces the control actions necessary (resource redistribution) to correct faulty system states. Managers use heuristic knowledge, that is, the decision-making process requires a specialised knowledge which is a result of the managers' experience. This is aggravated due to the huge expenses a wrong decision can lead to in the health care context. These problems require the use of artificial intelligence techniques for proper modelling. We have used knowledge-based systems [1], and the use of multi-agent systems [2] is currently on the rise.

The dynamic of the hospital, that is, its evolution in time, is determined by the flow of patients into the system and the resource distribution. The specific resources needed depend on the pathologies exhibited by the patients. A lack of resources or an increase in the number of patients with the same pathology will result in an inadequate state for the hospital. The symptom that the hospital's performance is below a desired quality factor value is detected by an increase in patient waiting times above a reference threshold. When that happens, the hospital administrator must maintain the best resource distribution possible to minimise the existing problem. Implementing a software program that simulates patient flow through the hospital would be a great aid in the decision-making process, as it would allow the administrator to experiment with different actions and analyse their effects.

5.2.1 Patient Flow Model

Subsystem 1 in Fig. 5.1 represents the dynamic model of the hospital, formed by an inter-related group of subsystems. To determine its component systems, we observe the flow of individuals through the hospital; that is, the subsystems are the different units necessary to treat the patients. In the schematic shown in Fig. 5.2, we can see how the patients move through the different hospital units.

This subsystem's uncontrolled input is the arrival of patients at hospital that is produced when a health centre refers a patient to the hospital. Depending on the type of pathology presented, the patient will need either medical or surgical intervention.

The first stop upon entering a hospital is Admissions (medical or surgical), where the patient's medical history is recorded and, after a physical examination, a determination is made of the necessary diagnostic tests and follow-up exams that must be made by Ancillary Services.

The different medical tests (analyses, radiographs, electroencephalographs, etc.) are performed in Ancillary Services, which has the characteristic of not having its own patients, that is, they do not treat patients with specific pathologies; rather, they attend patients of other hospital services (medical and surgical).

When the patient has received the results of his tests, he is directed back to Admissions, but the process now becomes progressive, where the results of the tests are studied and a course of treatment for the patient is developed, or, if needed, additional tests are scheduled. The patient must consult with his attending physician when the treatment is finished or when he obtains the results of the new tests, and he will remain in this loop until he is sent home or admitted to a ward.

If the patient is accepted due to a medical pathology, he must go through Admissions to determine when he can be admitted to the appropriate ward. After treatment, the patient is discharged and leaves hospital. If he is determined to have a chronic illness, he is transferred to a chronic care centre.

If the patient is admitted for surgery, he goes through Surgical Admissions to schedule his admission to the preoperative ward. He waits for a certain amount of time there while the necessary tests are performed to prepare him for surgery. Then, after the necessary time spent in the operating room, he goes to the PACU (post-

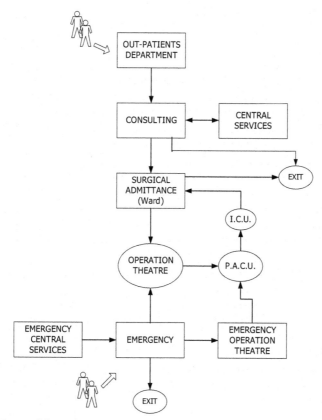

Fig. 5.2 Scheme of the patient flow in a hospital

anaesthesia care unit). From the PACU the patient goes to the postoperative ward until he completely recovers, after which he is discharged from hospital.

If at any time during his hospital stay the patient's condition worsens, he is admitted to the ICU (intensive care unit). When he recovers, he is returned to the appropriate ward to follow the treatment previously prescribed.

A patient may also arrive in hospital via the emergency room. In that case the necessary tests are performed on the patient (some of them carried out in the emergency room, others in Ancillary Services), and after specifying a treatment plan, he is admitted. If the patient has to be admitted to a ward, he goes through the emergency room Admissions, from which he is redirected to the appropriate surgical or medical ward, following the same pattern as other admitted patients. A third possible case arises when the patient is operated on in the emergency room, after which he goes to the PACU, and after his post-anaesthetic recovery, on to the appropriate ward to follow the usual path for patients of that ward.

5.3 Verification and Validation

For a model to be useful and for the experiments performed on it to be conclusive, we must have confidence in its predictive ability and in the results. Such confidence is obtained through a verification and validation (V&V) process [3].

Validating a model is not the same as verifying it. The validation consists of showing that the model is an accurate representation of reality. The verification implies testing the consistency of the design (adaptation to the problem of the modelling methodologies used, the algorithms, the software, etc.), that is, we must show that the model works as stated.

Validating a model is not an easy task since there is no one procedure or algorithm that indicates the steps necessary to accomplish it. What is more, since real systems are not completely known, and since models are never an exact representation of reality, the validation becomes more difficult still. The techniques used for assuring the degree of fidelity of a model or the level of accuracy with which it represents reality are usually the following:

1. Comparing the simulation results with historical data generated by the real system (retrospective validation).
2. Using the simulation to predict results, and then comparing these with the results generated by the system during some future time period (predictive validation).

Let us consider a process of V&V applied to the simulation of a simple hospital model. The hospital provided us with historical data (2003–2005) on patient arrivals in the operating theatre and on the length of the different procedures performed. We used a surgical process to retrospectively validate this simulation. Specifically, we focused on the ophthalmology department. Data for this same period related to the use of operating theatre resources by the ophthalmology staff has been also provided by the hospital. This simulation scenario has to maintain a steady state (a patient queue of zero), since we are simulating the arrival of patients in the operating theatre, their treatment and their egress. We did not consider the waiting list.

In order to model and simulate ophthalmology operating theatres, we need to be able to generate numbers at random that correspond with the arrival of patients and with the duration of the services. Given that we have the retrospective data we want to model, all that remains is to fit a theoretical distribution function.

The data provided regarding the duration of the services were not identically distributed. This forced us to divide them according to the diagnostic code for each operation. For each diagnostic code, we fitted the theoretical distribution function that most closely matched it. To guarantee a good fit we see if the hypothetical tests generate adequate values. Figure 5.3 shows how the adjustment was done by using the Arena simulation program's input analyser. Some other diagnostic codes, their service times and their arrival functions are shown in Table 5.1. The daily number of new arrivals was modelled with a Poisson function. Since on some days no new patients arrive, it was better to work with an arrival frequency rather than with a daily arrival rate. For this we used an exponential with a mean that was the inverse of the Poisson mean. The arrival of patients can be safely assumed to happen at the beginning of the day, since observed variables are measured at the end of the day.

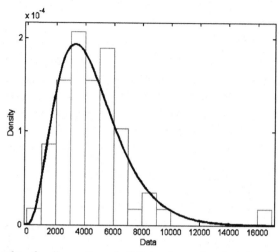

Fig. 5.3 Process time for diagnostic code 871 (ocular perforation) fitted to a Gamma with parameters $a = 3.753$ and $b = 1,203.48$

Table 5.1 Sample data fits (2003–2005) related to patient arrivals at ophthalmology operating theatres and service times

Diagnostic code	Sample size	Service time	Patient arrivals
216	119	300+3,600 * BETA(1.41, 3.18)	−0.5+LOGN(0.577, 0.205)
362	52	600+8,100 * BETA(0.695, 1.09)	POIS(0.0392)
366	1,423	600+10,800 * BETA(3.06, 7.63)	−0.5+9 * BETA(0.538, 2.88)
372	649	300+LOGN(1220, 1070)	−0.5+10 * BETA(0.388, 4.31)
373	98	600+WEIB(1590, 1.04)	−0.5+ERLA(0.0838, 7)
374	270	600+10,800 * BETA(1.29, 3.43)	−0.5+LOGN(0.68, 0.359)
375	59	1,200+10,500 * BETA(1.23, 1.7)	POIS(0.0484)

Data fit was validated by using simulation and checking that results yielded zero or near-zero queues. Thus, we can state that the model is an accurate representation of the structures (elements and inter-relationships) that exist in an ophthalmology operating theatre.

5.4 Experimentation

The hospital patient flow simulation lets us see how the hospital evolves over time. To this end, we have to convert the patient flow diagram (Fig. 5.2) into an algorithmic model (a program) that reveals the state of the different queues for specific situations.

The subsystem under consideration is a system of discrete events that changes with each new event. Such a system is not defined by a state equation; rather, an algorithm or flow chart must be devised to describe the evolution of the system as the events take place [4, 5]. To simulate these dynamics, we make use of a process-oriented methodology. A process is a time-ordered sequence of inter-related events separated in time. This sequence describes the passage of an entity (individual, data, etc.) through the system. To use an approach specific to the process, the sequence of steps (events or series of events) must be defined for each transaction taking place in the system, hence the name given to this simulation: a microsimulation. This sequence of steps can be represented with a flowchart. The objects needed by a transaction on its way through the system are called resources [6, 7].

A process-oriented, or process-based, simulation is widely used to simulate social systems (e.g. a hospital). Each part of the system obeys a common set of rules which govern the behaviour of each individual during their life cycle (arrival at hospital, first consultation, follow-up consultations, and so on). When these rules are defined in a computer, we are creating an artificial instance of a component. In a process-based simulation, this instance is called a process.

Each one of the processes uses resources. The simulation often makes use of resource competition. The entities have to wait in line before accessing the resources (for example, patients waiting to use an X-ray machine). Resource acquisition is the only method that allows processes to communicate. With a process-oriented model, the processes exchange or compete for resources. This limits the type of communication possible between the entities.

In the case of a hospital, we have a social system where the set of patients represents the entities. These entities communicate by way of the competition for resources (material or human). The program simulates the behaviour of each entity with time in a competitive environment. The programming languages best suited to this type of simulation are those which allow for the simultaneous execution of various instances of the same object (which may be executed differently depending on where they are in the simulation timeline), as well as for communication between them.

Object-oriented programming [8] is ideally suited to simulate these kinds of systems, since an object's state is defined by a combination of variables, and its behaviour by a set of methods. The fundamental data structures in object-oriented programming are the different object classes. So, to describe a process-based system, each of its elements must be modelled using an object which is an instance of one of the possible classes that exist in the system. Each particular instance of an object has its own state, even if the generic definition of its behaviour is set by the class to which it belongs.

In this chapter we are going to use the Java Discrete-Event Simulation System SIGHOS (from the Spanish initials for Hospital Resource Management Intelligent Simulation). This library allows for a process-oriented approach in modelling the system. That is, the system is characterised by elements which proceed along stages which differ according to the state of the element. In each stage, the elements perform different activities. To accomplish this, they use the system's own resources. This means that they wait for a resource to be available and they retain it as long as necessary to carry out the activity. The control over resource availability is through a timetable. This lets us specify when each resource will be available, as well as its role within the system. For example, in a hospital, the patients use one service and then another, making use of the hospital's resources. Some of these resources, like the doctors, perform functions in different units. Keeping in mind the characteristics of the problem, there are only two circumstances that affect the simulation timeline:

- An element engages in a voluntary waiting period (for example to simulate performing an activity once the necessary resources are obtained).
- An element tries to perform an activity and is forced to wait (for an unspecified time) for the necessary resources to become available.

The SIGHOS library gives the user the necessary mechanisms for representing the passage of time in the simulation. This is accomplished with a simulation time manager which is constantly deciding which system elements are active at any given time in the simulation, and which are on standby. The elements which must be defined in the library are:

- The elements moving through the system, carrying out activities and taking up simulation time, whether it be performing a process or waiting for resources
- The system resources necessary for the elements to carry out their activities

In addition, this library has an XML interface which allows one to model and simulate organisations without any knowledge of the Java language [9].

All the files needed to perform these experiments can be downloaded from (http://simull.isaatc.ull.es/bookchapter), including an adapted version of the SIGHOS library (*xmlsimulator.jar*). Each experiment includes the name of the model and experiment xml files which should be used:

- http://simull.isaatc.ull.es/bookchapter/model#.xml
- http://simull.isaatc.ull.es/bookchapter/exp#.xml

5.4.1 Implementation of a Gynaecological Consultation: Sequential Flow

The specification of the hospital dynamics is accomplished in two phases. In the first, the hospital is considered as consisting of one sole department. Once the resulting program is verified and validated, the simulation can be generalised to incorporate as many departments as exist in the actual hospital.

The first model using this verification process is a simplification of the arrival of patients at the gynaecology ward. Based on the results of this simulated example, we will be able to fit the model parameters to achieve the expected results. The timetables of the resources involved will be modified until patient waiting times are reduced to within desired limits.

A gynaecology ward is divided into two specialties: functional and oncological. When a patient arrives at hospital, she is initially treated for the pathology she exhibits. The length of the first consultation is 20 minutes, with a deviation of ±8 minutes for functional and 15 ± 5 for oncological. Once the first consultation is made, the patient has to undergo two more consultations, with a service time that follows a Normal(15, 5) for functional and Normal(10, 5) for oncological. The resources needed to carry out each consultation are an examination room and a doctor in each specialty.

In Figs 5.4a–c, we can see the XML code for implementing the previously described model. First (Fig. 5.4a) the simulation time unit is defined. For this example we use a minute since it is the smallest unit by which the simulation can advance (the activities last on the order of several minutes). The next step consists of describing the different types of resources: medical (functional and oncological) and material (functional and oncological examination rooms). Then, a timetable is defined which lists the instances of each resource. For example, there is a functional gynaecologist whose work day begins at 8 a.m. every day and lasts for 6 hours.

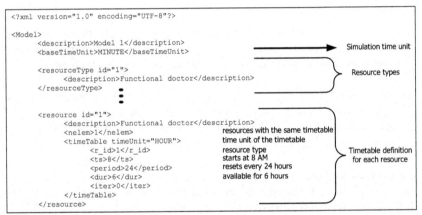

Fig. 5.4a XML description of patient gynaecological consultations with sequential flow (1/3)

Once the model's resources have been described, we define the activities performed by the patients (Fig. 5.4b): first and follow-up functional and oncological visits. For each activity, a duration and a list of necessary resources is specified.

Finally, we describe the patient flow for the different parts of the system (Fig. 5.4c). For the case under consideration, the patient carries out a sequence of activities consisting of an initial and two follow-up visits for the corresponding pathology.

Before being able to simulate the proposed model, the experiment parameters must be defined (Fig. 5.5). This is done by defining the number of repetitions to be executed and the number of days to be simulated. If we wish to recover a previous simulation state, we specify the starting and ending points of said simulation. In

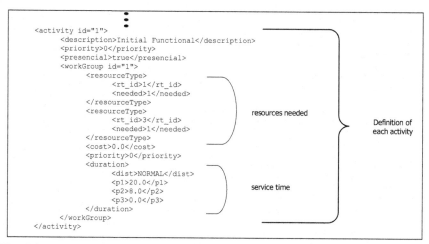

Fig. 5.4b XML description of patient gynaecological consultations with sequential flow (2/3)

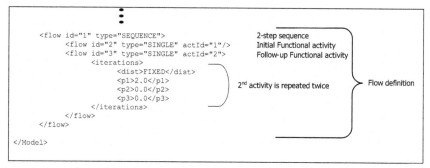

Fig. 5.4c XML description of patient gynaecological consultations with sequential flow (3/3)

Fig. 5.5 we simulate 30 days, resuming a previous 15-day simulation which had a zero initial state.

Next we present the results of the different experiments. They show the progression of the queues for different resource allocations, assuming an arrival rate of 20 new patients per day for each pathology.

Experiment 0 (*model0.xml, exp0.xml*) started from a zero initial state, that is, the hospital is empty and 15 days are simulated with a daily arrival rate of 20 patients for each pathology. The resources used are shown in Table 5.2 and the queue progressions are seen in Fig. 5.6.

We see how the line for the initial oncological consultation stays at an acceptable level. This is because this activity has more resources than the functional one. Additionally, its duration is shorter. The question, then, is: Why is the line for follow-up oncological consultations outside the limits if it uses the same resources as the first? The answer is that for each initial consultation, two follow-up consultations are

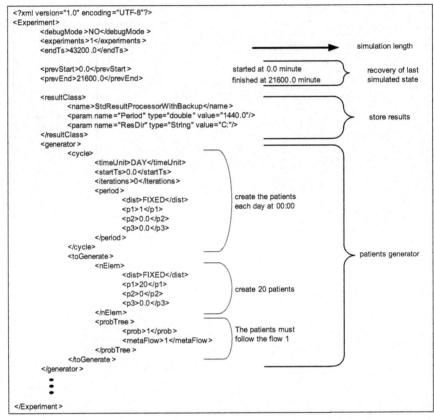

Fig. 5.5 XML description of the sequential flow model experiment performed on gynaecology patients

Table 5.2 Experiment 0: resources

Resource type	Amount	Availability (hours/day)
Functional (exam room + doctor)	1	6
Oncological (exam room + doctor)	1	8

created. This means that the initial consultation is acting as a source of patients for follow-up visits, and if all the patients have an initial consultation, then the queue for follow-up visits will gradually increase. Through this example we can see the existing inter-relationships between the different hospital services, where increasing the resources of one activity is not necessarily what is best for the hospital, since excessive waiting times may result elsewhere.

As for functional pathology, we see from Experiment 0 that the follow-up line does not increase. While this might be indicative of a properly functioning hospital,

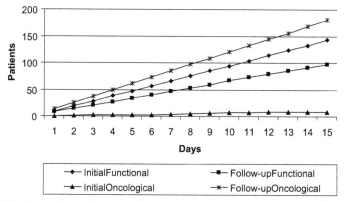

Fig. 5.6 Experiment 0: zero initial state

what is actually happening is that there are not enough resources to accommodate the patients at the initial consultation, resulting in a bottleneck which prevents the patients from arriving at the following activity.

Experiment 1 (*model1.xml*, *exp1.xml*) (Table 5.3 and Fig. 5.7) starts off where Experiment 0 ended but with increased resources. We see that it follows the same

Table 5.3 Experiment 1: resources

Resource type	Amount	Availability (hours/day)
Functional (exam room + doctor)	1	6
Functional (doctor)	1	6
Oncological (exam room + doctor)	1	8
Oncological (exam room)	1	8

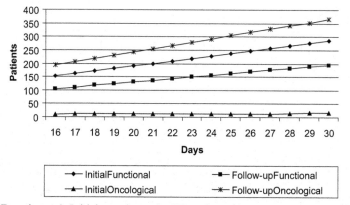

Fig. 5.7 Experiment 1: Initial state is result of Experiment 0

pattern, the only queue remaining at zero being for initial oncological. That is because the increase in resources does not result in more patients being seen, since in one case the number of doctors was increased but not that of examination rooms, resulting in an idle doctor with nowhere to examine his patients. In the other case, the number of rooms was increased but not that of doctors. These results show that an increase in the amount of money spent on more resources does not directly result in improved patient care. Due to the inter-relationships between resources, decisions involving which resources to increase and by how much should be made with the aid of a modelling tool.

In Experiment 2 (*model2.xml*, *exp1.xml*) (Table 5.4 and Fig. 5.8) the set of resources has been increased so as to attend to more patients in all of the activities. We see the resulting shortening of queues. For the case of functional follow-up visits, the tendency is for lines to increase since for every patient seen in the initial functional, two patients are generated for follow-up visits. A decrease in one queue means an increase in the other.

In Experiment 3 (*model3.xml*, *exp1.xml*) (Table 5.5 and Fig. 5.9) the resources have been increased so as to reduce all queues to zero. In many situations, this is not the best policy since the increase in quality is made at great expense to the hospital.

Frequently, the objective is to maintain waiting times within established limits, as in Experiment 4 (*model4.xml*, *exp4.xml*) (Table 5.6 and Fig. 5.10).

Table 5.4 Experiment 2: resources

Resource type	Amount	Availability (hours/day)
Functional (exam room + doctor)	1	6
Functional (exam room + doctor)	1	4
Oncological (exam room + doctor)	1	8
Oncological (exam room + doctor)	1	4

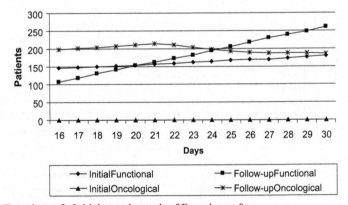

Fig. 5.8 Experiment 2: Initial state is result of Experiment 0

Table 5.5 Experiment 3: resources

Resource type	Amount	Availability (hours/day)
Functional (exam room + doctor)	4	6
Functional (exam room + doctor)	1	4
Oncological (exam room + doctor)	4	8
Oncological (exam room + doctor)	1	4

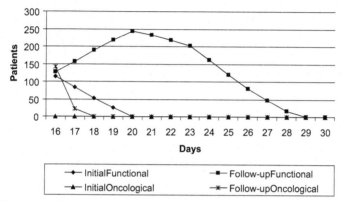

Fig. 5.9 Experiment 3: Initial state is result of Experiment 0

Table 5.6 Experiment 4: resources

Resource type	Amount	Availability (hours/day)
Functional (exam room + doctor)	2	6
Functional (exam room + doctor)	1	4
Oncological (exam room + doctor)	1	8
Oncological (exam room + doctor)	1	4

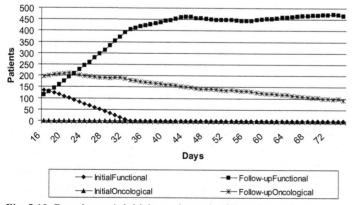

Fig. 5.10 Experiment 4: initial state is result of Experiment 0

5.4.2 Implementation of Gynaecological Visits and Diagnostic Tests: Combined Sequential–Simultaneous Flow

Next we model and simulate patient flow through the gynaecology ward, taking into account the diagnostic tests made before the follow-up visit (Fig. 5.11). As previously stated, the gynaecology department is divided into two specialties: functional and oncological. When a patient arrives at hospital presenting the symptoms of either pathology, she or he is seen by the appropriate specialist. Blood tests are then performed with a service time Normal (8, 3) and an ultrasound with time Normal (20, 10) before returning for the follow-up visit. These last steps are repeated between two and four times. The tests (bloodwork and ultrasound) are requested simultaneously since the order in which they are performed is not important. Of course, the patient cannot have both tests done at the same time. The service times

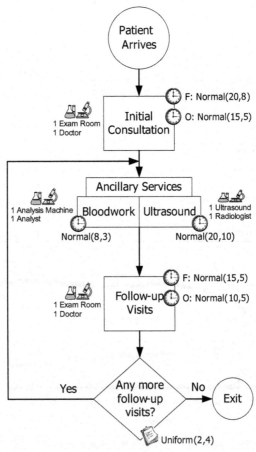

Fig. 5.11 Process for a gynaecology patient

for each consultation are as specified in the previous section. Each pathology has its own specific exam room and doctor. The blood work requires a technician and an analysis machine, the ultrasound an ultrasound machine and a radiologist.

The XML code changes with respect to the model of Sect. 5.4.1 in that new resources have to be introduced (analysis machine, technician, ultrasound machine and radiologist), the two activities for the diagnostic tests have to be defined, and finally the patient flow has to be modified (Fig. 5.12).

The next set of experiments assumes a daily arrival rate of 40 new patients for each pathology. Figure 5.13 shows the results for a system with a zero initial state using resources from Table 5.7 (*model5.xml, exp5.xml*). The graph indicates that all the queues are around zero, except for the analysis queue, and the first and follow-up functional consultations. One possible corrective action would be to take away resources from the oncological consultations and radiology, which are operating below capacity. That is to say, oncological consultation and radiology financing is reduced, thus limiting the hours these resources are available. Such remaining financial resources can be used to raise the availability of other resources.

However, Fig. 5.14 shows that only by adding more resources for analysis, the queues for oncological activities increase even without reducing their resources as stated in Table 5.8 (*model6.xml, exp6.xml*). This is because in Experiment 5, the analysis activity was acting like a sink; that is, since there were not enough resources to accommodate all the patients in analysis, and since without getting those tests done they could not continue in the flow, patients piled up in the analysis queue, making it appear as though resources abounded in the other activities.

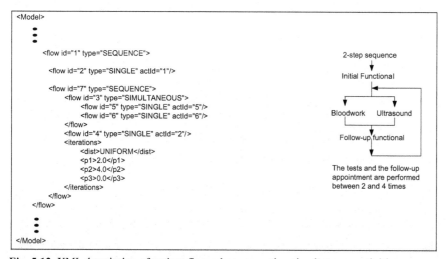

Fig. 5.12 XML description of patient flow when requesting simultaneous activities

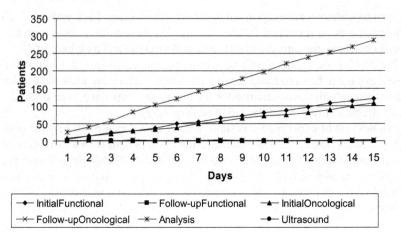

Fig. 5.13 Experiment 5: consultation and diagnostic test simulation from a zero initial state

Table 5.7 Experiment 5: resources

Resource type	Amount	Availability (hours/day)
Functional (exam room + doctor)	2	8
Analysis (analyst + equipment)	1	6
Oncological (exam room + doctor)	2	6
Ultrasound (radiologist + equipment)	4	8

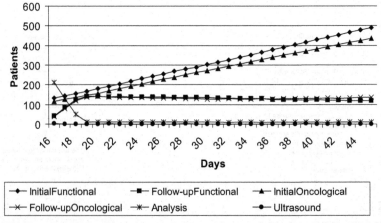

Fig. 5.14 Experiment 6: consultation and diagnostic test simulation with the initial state from the result of Experiment 5

Table 5.8 Experiment 6: resources

Resource type	Amount	Availability (hours/day)
Functional (exam room + doctor)	2	8
Analysis (analyst + equipment)	3	6
Oncological (exam room + doctor)	2	6
Ultrasound (radiologist + equipment)	4	8

5.5 Conclusions

In this chapter we have described the hospital management problem, showing how good management of hospital resources is necessary to achieve the quality-of-life levels we all desire. The complexity of a hospital system makes managing its resources a non-trivial matter. It is a system made up of many parts (patients and resources) and many inter-relationships (departments competing for resources, resources with overlapping roles in time, patients performing activities in various departments, etc.). All of this makes the use of modelling and simulation techniques necessary in order to have a detailed knowledge of the status and to be able to act accordingly (to correct deficiencies).

The different experiments performed have shown that a simulation using discrete-event systems (DES) is a viable methodology for dealing with the flow of patients through the various departments of a hospital. Furthermore, we introduced the DES tool called SIGHOS, which lets us simulate the hospital via a process-based approach. Finally, we also remark upon the XML interface included in the library which allows for modelling a system without any arduous programming.

5.6 Questions

1. In the examples we assume that all the patients arrive at the hospital at 00.00 every day. Is this approach equivalent to uniformly distributing the arrival of patients during the availability period of the resources?
2. As you can see from the previous question, we have introduced two different approaches to model the arrival of patients. What scenarios do you think are best suited for each approach?
3. Can this type of simulation be used to analyse other systems?

Acknowledgments This work is supported by a project (reference DPI2006-01803) from the Ministry of Science and Technology with FEDER funds. Iván Castilla is supported by an FPU grant (ref. AP2005-2506) from the Spanish Ministry of Education and Science.

References

[1] Moreno L, Aguilar RM, Piñeiro JD et al (2001) Using KADS methodology in a simulation assisted knowledge based system: application to hospital management. Expert Syst Appl 20:235–249

[2] Aguilar RM, Castilla I, Muñoz V et al (2005) Predictive simulation for multiagent resource distribution: an application in hospital management. In: Proceedings international Mediterranean modelling multiconference, EMSS 2005

[3 Sargent RG (1998) Verification and validation of simulation models. In: Proceedings 1998 winter simulation conference

[4] Moroza A (2006) Simulation applications in health care. M S Thesis, Faculty of computer Science and Information Technology, Riga Technical University, Riga, Latvia

[5] Pidd M (2004) Computer simulation in management science. John Wiley, Chichester

[6] Banks J, Carson JS, Nelson BL, Nicol DM (2000) Discrete-event system simulation. Prentice Hall, Upper Saddle River, NJ, pp 367–397

[7] Law AM, Kelton WD (1991) Simulation modeling and analysis. McGraw-Hill, New York

[8] Jacobson I, Christensen M, Jonsson P, Overgaard G (1992) Object-oriented software engineering: a use case driven approach. Addison-Wesley, New York

[9] Aguilar RM, Muñoz R, Castilla I et al (2006) An interface to integrate workflow diagrams into discrete-event simulation. In: Proceedings international Mediterranean modelling multiconference, EMSS 2006

Chapter 6
Supply Chain Cyclic Planning and Optimisation

G. Merkuryeva and L. Napalkova

Abstract This case study analyses different simulation-based optimisation methods of multi-echelon supply chain planning in the maturity phase of the product life cycle. Some standard optimisation software add-ons as well as the proposed model in the case study are used to solve the problem. A supply chain generic network is employed as an application system. Several optimisation scenarios are introduced in order to analyse and compare abilities of different optimisation methods and tools. A hybrid genetic-response surface-based linear search algorithm is introduced to enhance the solution of multi-echelon cyclic planning and optimisation problems and generate the optimal cyclic plan.

6.1 Introduction

In this case study, we propose analysing a generic network to cope with a manufacturing supply chain planning and optimisation problem for products in the maturity phase of their life cycle. The generic network contains 42 stock points and 41 processes that describe storages/warehouses, purchasing, production, packing, distribution and transportation processes, correspondingly. The planning manager aims to decrease the supply chain total cost while satisfying end-customers' demands and trying to increase customer service levels. A simulation study is proposed to analyse abilities of different simulation-based optimisation methods and tools to solve the problem; introduce a hybrid genetic-response surface-based linear search algorithm; and generate an optimal cyclic plan that defines the optimal lengths of

Galina Merkuryeva and Liana Napalkova
Riga Technical University, Latvia
gm@itl.rtu.lv, liana@itl.rtu.lv

Y. Merkuryev et al. (eds.), *Simulation-Based Case Studies in Logistics*
© Springer 2009

the planning cycles and quantities to be ordered or produced for each mature product in the generic network.

To manage supply chains, two different approaches are used in practice. The single-echelon approach splits a multi-level supply chain into separate stages where a stage or facility is managed independently. The so-called multi-echelon approach [1] considers planning and managing the supply chain as a whole and thus allows optimising the global supply chain performance. A multi-echelon environment considers multiple processes (e.g. purchasing, production, packing and transportation) and multiple stock points (storage and warehouse). Planning and optimisation of production and inventories in multi-echelon environments is the scope of this case study.

Cyclic and non-cyclic planning policies are applied in multi-echelon environments. The underlying idea of cyclic planning is to use cyclic schedules for long-term planning at each echelon and to synchronise them with one-another [1]. Every process in the supply chain, whether it is a purchasing, production, packing or transportation, is planned on a repetitive, 'cyclic' basis, and the process cycles are synchronised and fit together (see Fig. 6.1, where a cycle at each echelon is represented by a 'planning wheel'). Cyclic schedules are preferable for the product stable demand. When demands are dynamic, flexibility in spacing production periods could result in a lower total cost of non-cyclical schedules. However, the real-life performance of a specific planning policy may differ from the theoretical one.

Cyclic schedules offer practical benefits in terms of easy planning and control, and reduce administrative costs for monitoring planning policy. Cyclic long-term benefits could result [2] in reduction of safety stock buffers between echelons, time and cost of material handling, expected order and production times and costs, etc.

This case study is a simplification of a more extensive research study. Nevertheless, the application problem is simplified, since our main goal is to present and discuss the simulation-based optimisation methodology. Here, a simulation model is built to estimate system performance, while an optimisation algorithm uses the responses generated by the simulation model to control the optimisation process. All e-educational materials supporting this case study are available at http://www.itl.rtu.lv/Case_studies_Chapter_6/.

6.2 Problem Definition

In this section we will formulate the problem. In particular, we will introduce assumptions that define the scope of the problem, describe the main objective func-

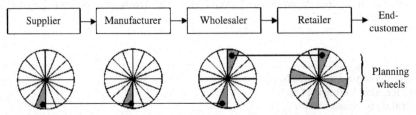

Fig. 6.1 Synchronisation of 'planning wheels' in a multi-echelon supply chain

tions, define decision variables and constraints, and finally will perform the problem express analysis.

6.2.1 Assumptions

The following assumptions are introduced to define the scope of the problem:
- Demand is considered to be uncertain; while predicting the demand average value, its variations are estimated by a standard deviation of the demand per period.
- Lead times of the processes are known and constant.
- Lot sizes of the products are variable.
- Capacities are limited.
- Demand is considered to be independent only for customised products.
- Backorders are delivered in full.
- Costs: fixed production and ordering costs, and linear inventory holding costs are assumed.
- Planning is performed for a finite planning horizon.

6.2.2 Objective Functions

We define two main objective functions in the problem. The first one is aimed towards minimising the average total cost, which includes the sum of inventory, production and reordering costs:

$$\text{Minimise } TC = \sum_{t=1}^{T} \left(\sum_{j=1}^{J} CP_{jt} + \sum_{i=1}^{I} CO_{it} + \sum_{i=1}^{I} CH_{it} \right), \tag{6.1}$$

where TC denotes the total cost, CP_{jt} denotes production cost of process j per period t, CO_{it} is ordering cost at stock point i per period t, and CH_{it} is inventory holding cost at stock point i per period t; I and J correspond to the number of stock points and processes, and T defines the number of periods in the planning horizon.

To avoid unconstrained minimisation of the total cost and satisfy customer service requirements, we introduce the second objective. It is aimed towards maximising the average order fill rate FR, which is defined as the percentage of end-customers' orders filled from the available inventory:

$$\text{Maximise } FR = \frac{100 * \sum_{t=1}^{T} \sum_{i=1}^{I} \sum_{k=1}^{K} QC_{ikt}}{\sum_{t=1}^{T} \sum_{i=1}^{I} \sum_{k=1}^{K} D_{kit}}, \tag{6.2}$$

where QC_{ikt} is the sum of orders delivered by stock point i to end-customer k in time period t, D_{kit} is the actual demand of end-customer k to stock point i in time period t.

6.2.3   Decision Variables

The parameters of a multi-echelon cyclic plan identify multiple decision variables to be optimised in the problem. They are: replenishment cycles Cy_i and order-up-to levels S_i defined at each stock point i on the network. These variables determine the reorder period and quantity to be ordered or produced for each mature product in the network and interpreted as discrete and continuous variables, correspondingly.

The number of decision variables increases with the number of stock points. As a result, a large number of decision variables in practice could make conducting interactive optimisation experiments difficult. Moreover, metric scales of decision variables have a very different range of possible values. During simulation-based optimisation experiments, 'order up-to-levels' type variables are calibrated with a discrete step size defined by 10 product units in the case study.

6.2.4   Constraints

The problem constraints include decision variable constraints, storage and fill rate constraints, and specific cyclical replenishment constraints:
1. *Decision variable constraints.* Search spaces for decision variables, i.e. replen-ishment cycles Cy_i and order-up-to levels S_i, can be limited by a lower (min) and upper (max) bound: $\cdot \forall i,\ Cy_{min} \leq Cy_i \leq Cy_{max}$, $\cdot \forall i,\ S_{min} \leq S_i \leq S_{max}$.
2. *Fill rate constraint* expresses a minimum fill rate value FR_{min} that has to be satis-fied for end-customers in a supply chain, i.e. $\cdot FR \geq FR_{min}$.
3. *Storage capacity constraints* express the level of stock that can be managed by a stage. On-hand stock H_{it} at the end of period t is not allowed to exceed the capac-ity of inventory buffer CAP_i: $\forall i,t\ H_{it} \leq CAP_i$.
4. *Cyclical replenishment constraints* are introduced to define a cyclic policy type. If replenishment cycles are constrained by *integer-ratio policy,* the reorder inter-val at any stage is the only integer. If replenishment cycles are constrained by *power-of-two policy*, the cycles are integers and are defined by power of two multiples of a basic period q, i.e. for each stock point m, $Cy_m = 2^p q$, where m and q are non-negative numbers. For example, if $q = 7$, then cycles could be equal to 7, 14, 28, 56, etc.

6.2.5   Express Analysis

The problem express analysis shows that the application of the MILP analytical model (mixed integer linear programming [2]) in multi-echelon cyclic planning is limited by assumptions of a constant demand and lead times. These assumptions significantly decrease the complexity of the problem, but could still be considered useful for mature products.

The problem formulation as a stochastic dynamic programming model does not have an efficient analytic or heuristic algorithm, which could find the optimal solution. Moreover, in some cases mathematical simplifications could result in sub-optimal solutions.

The stochastic discrete-event simulation technology does not require a rigid structure for the analytical model and provides an experimental approach to supply chain analysis and optimisation. It allows the analyst to dynamically model complex interactions between systems entities; to model processes that contain nonlinearities; to introduce variability of demand and multiple objectives into multi-echelon planning procedure; to introduce problem-specific constraints, for example, assign capacity constraints to inventory locations; and take into account constraints at the supply chain different echelons. Simulation is used to estimate the performance of the system for each set of decision variables' values, but it is not sufficient by itself to yield the desired quality of a simulated system.

In the following sections, we will present an optimisation approach to supply chain planning and optimisation based on using a discrete-event simulation model to evaluate the objective function values in the problem.

6.3   Simulation Model Description

A generic network simulation model itself is built as the process-oriented one. It models storage and processing of raw materials, transportation of semi-finished products to direct customers or to a plant where other components are added to make different products. Then finished products are shipped to different types of end-customer sites.

The supply chain network is represented by two types of atomic elements: stock points and processes. The *stock point* defines any buffer or storage where output products of the process are stored. The *processes* correspond to transformations of a set of input products to a set of output products, such as production, packing and transportation operations. Stock points and processes are graphically represented by triangles and rectangles, correspondingly (see Fig. 6.2). Any process with a stock point connected with a directed arc defines a stage. A set of stages that belong to the same network level creates an echelon. Input parameters, decision variables and constraints are assigned to the atomic elements. The network is supposed to have a one-directional flow of goods.

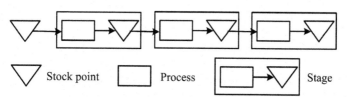

Fig. 6.2 Conceptual model of the supply chain linear network

In the network simulation model, the processing logic is defined for each stage in the network. It is initialised at the beginning of each period, when the end-customer demand is generated. The replenishment order is placed to the immediately preceding stage. Orders and backorders are delivered to the immediately succeeding stage. If on-hand stock is insufficient to fulfil this order, then the backorder is created. Only full backorders are processed in the model. At the end of a period, the decision about replenishment order is made according to the planning policy, and corresponding costs are calculated. The sequence of events processed in the simulation model is shown in Fig. 6.3.

The controllable inputs to the simulation model are associated with the problem decision variables, described in Sect. 6.1, and define replenishment cycles Cy_i and order-up-to levels S_i for each stock point i in the network. End-customer demand distribution parameters, i.e. an average demand and its standard deviation, identify the model uncontrollable inputs. Production, ordering and holding costs, process lead times and stock point capacities are regarded as model parameters. Finally, responses or performance measures of the simulation model are associated with objective functions (6.1), (6.2) and are estimated by their average values, i.e. by an average total cost $E(TC)$ and an average fill rate $E(FR)$.

The correspondent simulation model of the multi-echelon generic network has been automatically generated [3] in a ServiceModel Professional 7.0 simulation environment and consists of its standard elements, i.e. locations, entities, arrivals, etc. The model layout is presented in Fig. 6.4. The supply chain network is defined as an input in the format that allows automatic reading of it within the simulation optimisation experiments.

To verify the model generated within the simulation project, a chart-based tracing procedure was used. It will not be described here in detail. Let us mention the main points verified while tracing the model: tracing of sent and received orders, and checking when on-hand inventory is equal/is not equal to the inventory position.

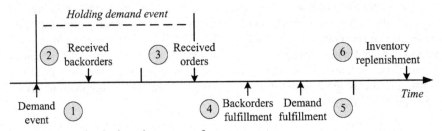

Fig. 6.3 Processing logic and sequence of events

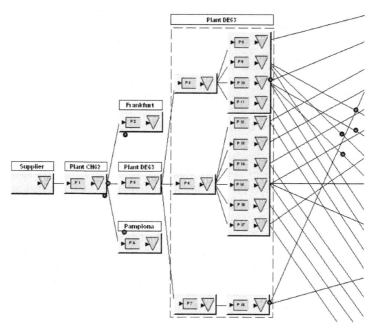

Fig. 6.4 Simulation model screenshot

6.4 Optimisation Methodology

In this section we will describe a simulation-based optimisation scheme and provide a review of methods and software used in this case. Finally, we will describe a hybrid simulation optimisation algorithm for solving the cyclic planning problem.

6.4.1 Simulation-Based Optimisation Scheme

The simulation-based optimisation approach has recently become a 'hot' technology in supply chain planning and management. It is aimed towards finding which of possibly many sets of decision variables leads to the optimal or near-optimal performance of the model by associating objective functions with simulation model performance measures. An optimisation algorithm that utilises a discrete-event simulation model runs simulation to collect observations and estimate system performance measures.

In the simulation optimisation procedure, the optimisation algorithm chooses values of decision variables and uses the responses generated by the simulation model to make decisions regarding the selection of the next potential solution. The general scheme of the simulation optimisation procedure is shown in Fig. 6.5.

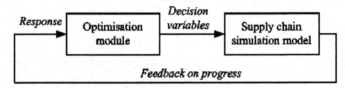

Fig. 6.5 Simulation optimisation scheme

To define initial values of decision variables, or a starting cyclic solution in simulation optimisation experiments, simplified analytical formulas that calculate parameters of a multi-echelon cyclic plan are used. These calculi are not explained in the text; they can be found in the Appendix.

To estimate the simulation model responses that are defined by the average total cost and the average fill rate, multiple runs with the simulation model are performed for each set of decision variables. The initial state of the simulation model and the length of simulation run are defined by the initial inventory levels and planning horizon, respectively.

6.4.2 Optimisation Methods and Software Add-On

Let us formulate the problem *requirements* to the simulation-based optimisation method in this case study. The method should introduce multiple objectives and avoid getting the local optima, should be able to deal with both discrete and continuous variables, should incorporate constraint handling techniques and should avoid getting non-stable optimal solutions that could not be properly implemented in practice. 'False' solutions related to stochastic perturbations of the objective function in conditions of a stochastic demand should also be neglected.

To handle *multiple objective functions*, one could use aggregating multiple objectives into a single objective. The main strength of this approach is a computational efficiency and simple implementation. The weakness is the difficulty in determining weights that reflect a relative importance of each objective [4]. Moreover, this approach can produce the only one optimal solution during a single experiment. However, in multi-objective optimisation problems, each objective function could have its optimal solution and none of them could be considered as better than any other with respect to all objective functions. In order to find a set of trade-off solutions, the concept of Pareto optimality is introduced. A solution $\mathbf{x}^*$ is Pareto-optimal if it is not worse than other solutions for all objectives and is better for at least one objective. The Pareto-optimal set defines a set of non-dominated solutions in the entire search space.

Based on the state-of-the-art in the area of simulation optimisation, four groups of *optimisation methods* could be identified: (1) stochastic approximation methods, (2) gradient-based methods, (3) response surface methods, (4) random search methods and metaheuristics. For other classifications we refer to Azadivar [5], Merkuryev and Visipkov [6], Carson and Maria [7], Fu and Glover [8].

In the context of formulated requirements, the last group of heuristic methods seems to be the most promising one. Among the heuristic techniques, population-based heuristics such as *evolutionary algorithms* are well suited for optimising cyclic plans in multi-echelon supply chains. Moreover, specific cyclical constraints that present the power-of-two replenishment policy could be easily satisfied in genetic algorithms (GAs) by using a binary alphabet to encode candidate solutions.

Different *multi-objective genetic algorithms,* while handling multiple objective functions, (1) convert multiple objective functions into a single one, or (2) use a separate population for each objective function, or (3) apply non-domination concepts. In particular, the multi-objective GAs based on non-domination concepts, i.e. non-dominated sorting genetic algorithm, niched-Pareto genetic algorithm or strength Pareto evolutionary algorithm, calculate individuals' fitness on the basis of the Pareto dominance principle and approximate the Pareto-optimal front without prioritising, scaling and weighting objective functions. For a detailed discussion of multi-objective evolutionary algorithms the interested reader is directed to [9].

None of these algorithms could work with both discrete and continuous decision variables as it is required in the optimisation problem statement. The hybridisation of several methods is a widely used approach to develop more powerful algorithms, which produce better solutions to the complex problems. The hybridisation approach is widely used by commercial optimisation tools, which are software add-on compatible with discrete-event simulation tools. In the case study, SimRunner® (an optimisation software add-on included in ServiceModel Optimisation Suite) and OptQuest® for ProModel (an optimisation software add-on for users of Pro-Model/MedModel/ServiceModel simulation tools) are used in Sect. 6.5 to introduce experimentation scenarios with specific cyclical constraints defined by the integer-ratio policy.

SimRunner® (http://www.promodel.com/products/simrunner) optimisation algorithm aggregates multiple objectives into a single objective; implements genetic algorithms and evolution strategies to explore a search space; and introduces neural network-based metamodels to estimate the model performance without performing simulation experiments, in order to decrease the time necessary to solve an optimisation problem.

OptQuest® for the ProModel (http://www.promodel.com/products/optquest) optimisation algorithm aggregates multiple objectives into a single objective; implements scatter search to explore the solution space with a smaller number of objective function evaluations; applies adaptive memory concepts to avoid the re-investigating of the solutions already evaluated; uses MILP models and simplex methods to handle constraints, and the already-mentioned neural network-based metamodels.

Another hybrid optimisation algorithm is introduced in Sect. 6.4.3; it integrates genetic algorithms based on non-domination concepts with response surface-based linear search methods. It is used in Sect. 6.5 in order to introduce simulation optimisation scenarios with specific cyclical constraints defined by the power-of-two policy and to analyse if simulation optimisation methods that are able to search Pareto-optimal solutions could lead to better solutions to the problem.

6.4.3 A Two-Phase Hybrid Optimisation Algorithm

The hybrid optimisation algorithm, aimed towards defining the optimal parameters of a multi-echelon cyclic plan, is based on the cooperative search of the *genetic algorithm* and *response surface-based method (RSM)*. While GA is well suited to solve combinatorial problems and is used to guide the search towards the Pareto optimal front, RSM-based linear search is appropriate to the improvement of GA solutions based on the local search approach and applicable for continuous decision variables.

The hybrid optimisation algorithm consists of the following phases: (1) simulation optimisation of replenishment cycles using an improved multi-objective genetic algorithm, and (2) simulation optimisation of order-up-to levels using the RSM-based linear search method.

6.4.3.1 Phase 1. Simulation Optimisation Using a GA

Simulation optimisation using an improved genetic algorithm integrates the non-dominated sorting genetic algorithm II [10], problem-specific constraints handling techniques and a discrete-event simulation model to optimise lengths of replenishment cycles in a multi-echelon cyclic plan. Here, a cycle is defined by power-of-two policy and codified by a binary string or sequence of genes. These genes are composed into a chromosome, which represents a set of replenishment cycles to be optimised. Each chromosome corresponds to a potential solution of the problem (see Fig. 6.6). The corresponding values of order-up-to levels are calculated using approximate analytical calculi (see Appendix).

Fitness of the chromosome is assigned based on two objective measures, i.e. the total cost and fill rate, obtained from simulation experiments. Here, the concepts of the non-dominated solution and domination depth are applied. For example, the solution represented by point A in Fig. 6.7 is better than the solution at point B, as it gives higher fill rate at lower total cost. This means that solution A is non-dom-

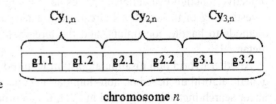

Fig. 6.6 Example of a chromosome and genes

Fig. 6.7 Illustration of the dominance relation

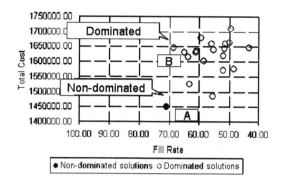

inated and belongs to the first non-dominated front (domination depth is 1), while
solution B is dominated.

To maintain the diversity of the solutions in the Pareto front, the crowding distance of each chromosome, which estimates density of the solutions surrounding
the current solution, is introduced. It is calculated based on the total cost and fill
rate normalised values. So each chromosome in the population has two attributes,
i.e. a domination depth, and a crowding distance. From two solutions the one with
the lower depth is preferable. If both solutions have the same depth value, then the
one with the larger crowding distance is preferable.

To avoid the loss of non-dominated solutions during the evolution process, the
so called $(\mu + \lambda)$-selection scheme [4] is used, where μ and λ assign parents and mating pool, respectively. Finally, a stopping criterion is defined based on the convergence speed towards the Pareto optimal curve. Figure 6.8 depicts a flow diagram of
this algorithm; it includes the following steps:

1. *Initialisation:* Generate an initial population P_N of replenishment cycles Cy_i randomly by using a uniform distribution to smoothly cover the search space and
   calculate corresponding values of order-up-to levels S_i. Define a lower bound N
   of the population size that guarantees both genetic diversity and reasonable processing time [11], i.e. $N = 1.65*2^{(0.2*L)}$, where L is the length of a chromosome.
2. *Solutions encoding:* Codify lengths of stock point replenishment cycles by the
   power p of base 2 using a binary string a_L. For example, the cycle $Cy_i=28$ is
   represented as $7*2^2$, where a basic period or a minimal length of a cycle is equal
   to 7 days. Then, the power $p=2$ of the base 2 is encoded to a binary string
   $a_2=<1, 0>$, i.e. $0*2^0+1*2^1=2$.
3. *Fitness assignment:* Estimate objective functions, i.e. the total cost and fill rate,
   for individual solutions or chromosomes in P_N through simulation experiments
   with a supply chain model. To assign fitness, first find non-dominated solutions,
   or chromosomes in the entire population P_N and assign them domination depth
   $r_n=1$. Temporally exclude non-dominated solutions from the population. Find
   new non-dominated solutions in the remaining population and assign them the
   domination depth $r_n=r_n+1$, etc. Finally, reorder individual solutions according
   to their domination depth.

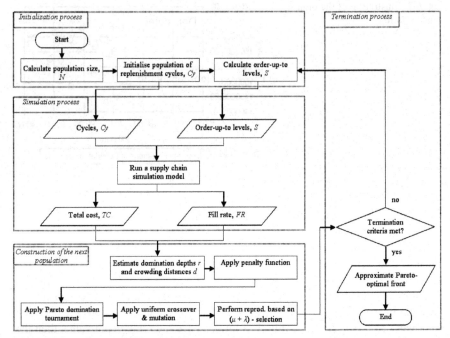

Fig. 6.8 Flow diagram of GA algorithm

4. *Constraints handling:* Apply a penalty function to infeasible solutions in the current population that have fill rates lower than the pre-defined threshold. To reduce the survival probability of these solutions, increase artificially total costs of the chromosomes by $TC*k$, where k is a multiplier coefficient that could be adjusted during the optimisation process. Here, FR_{min} is defined by 75% and $k=2$.

5. *Selection operation:* Calculate crowding distances for all chromosomes in the current population and fill the mating pool, i.e. select the pairs of individual solutions as parents in reproduction operation (or 'fill the mating pool') by using two-tournament selection scheme based on the attributes of chromosomes randomly selected from the current population.

6. *Reproduction operation:* Perform uniform crossover and one-point mutation operations to the mating pool in order to generate the offspring. Combine the offspring with the parents' population and update domination depths of chromosomes in the combined population. Include the first N solutions in the new population.

7. *Termination:* If the stopping criterion is met, terminate the search; otherwise, return to Step 3.

6.4.3.2 Phase 2. RSM-Based Linear Search

Phase 2 is aimed towards improving the cyclic planning solution received from a genetic algorithm in Phase 1 by adjusting analytically calculated order-up-to levels for stock points. It is based on the response surface-based methodology applied to a simulation optimisation problem [12], in which the total cost is introduced as simulation response and order-up-to levels as input factors. The RSM-based linear search algorithm presents a sequential procedure and is applied to all solutions from the Pareto-optimal front initially generated by the GA.

In iteration m a small experimentation region of input factors is described by the I-dimensional rectangle $[t_1^m, u_1^m] \times \ldots \times [t_I^m, u_I^m]$ with a central point ξ_i^m and a step c_i^m defined by:

$$\xi_i^m = \frac{t_i^m + u_i^m}{2}, \quad c_i^m = \frac{u_i^m - t_i^m}{2} \quad i = 1, \ldots, I \tag{6.3}$$

where t_i^m and u_i^m define a lower and an upper bound of input factor i.

In iteration m the procedure consists of the following steps :

1. *Local approximation of the response surface function.* Approximate the response surface function in a small region of input factors using the first-order regression metamodel, which describes the main effects of input factors. The central point for an initial region of experimentation is taken from the GA results received in Phase 1. To increase the numerical accuracy in estimation of regression coefficients, input factors are coded by formula (6.4) and the coded first-order model (6.5) with coded variables x_i^m (see Fig. 6.9) is introduced:

$$x_i^m = \frac{\xi_i^m - \xi_{i0}^m}{c_i^m}, \quad i = 1, \ldots, I, \tag{6.4}$$

$$y = b_0^m + \sum_{i=1}^{I} b_i^m x_i^m + \varepsilon^m, \tag{6.5}$$

where ε^m is a statistical error of a regression model. To estimate coefficients b_i^m from simulation experiments, the Plackett–Burman experimental design added by simulation experiments replicated in the central point could be applied. The template of this design is automatically generated in Minitab statistical software. In the case of 33 input factors it includes 36 experiments. Note that while regression coefficients of a metamodel are defined based on coded values of input factors, simulation experiments are performed for their natural values.

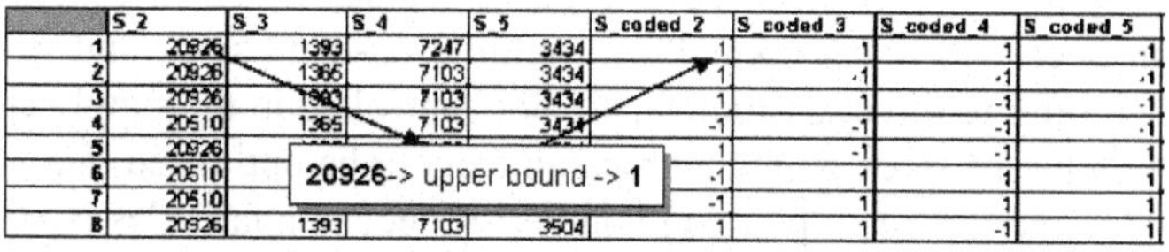

	S_2	S_3	S_4	S_5	S_coded_2	S_coded_3	S_coded_4	S_coded_5
1	20926	1393	7247	3434	1	1	1	-1
2	20926	1365	7103	3434	1	-1	-1	-1
3	20926	1393	7103	3434	1	1	-1	-1
4	20510	1365	7103	3434	-1	-1	-1	1
5	20926				-1	-1	-1	1
6	20510	20926-> upper bound -> 1			-1	1	1	1
7	20510				-1	1	1	1
8	20926	1393	7103	3504	1	1	-1	1

Fig. 6.9 Example of encoding procedure

2. *Checking the fit of a local metamodel.* Perform the lack-of-fit test using the
   p-values based on an ANOVA table in order to check the adequacy of a regression metamodel. A significant lack-of fit, i.e. $p > 0.05$, may be a result of factor
   interactions excluded from the model. If it occurs, it would be reasonable to
   decrease the size of the current experimental region in order to fit a first-order
   approximation model. Moreover, in the case of model adequacy, the direction
   of significant improvement for a simulation model response could not be easily
   found if estimates of regression coefficients received are quite small compared
   with an estimation error. In this case increasing the length of a simulation run or
   the number of replicate runs for each experimental point may guarantee statistical significance of the search direction in Step 3.
3. *Linear search in the steepest descent direction.* Perform a linear search within
   the local search space for order-up-to levels, in the steepest descent direction
   defined by a vector $(b_1^m, b_2^m, \ldots, b_I^m)$ starting from the central point, $b_1^m, b_2^m, \ldots,$
   b_I^m are coefficients of the simulation metamodel received in iteration m. The
   increments Δx_i^m, $i = 1, \ldots, I$ along the projection of the search direction are calculated for coded factors only for significant regression coefficients (with p-value
   < 0.05) taking account their main effects:

$$\Delta x_i^m = \frac{-b_i^m}{\max_i |b_i^m|}, \; i = 1, \ldots, I. \tag{6.6}$$

$$\Delta_i^m = \Delta x_i^m * c_i^m, \; i = 1, \ldots, I. \tag{6.7}$$

The next line search point in iteration m is calculated as follows:

$$\zeta_i^m = \zeta x_i^m + \Delta x_i^m, \; i = 1, \ldots, I. \tag{6.8}$$

The linear search in iteration m is terminated, if the simulation response value can
not be improved.

Finally, the Pareto-optimal front initially generated by a GA is updated including
solutions found in the RSM-based linear search procedure. Solutions received are
reordered according to their fitness values in the increasing sequence.

6.5 Experimentation

In this section we will introduce four simulation optimisation scenarios in order to
generate an optimal multi-echelon cyclic plan for a supply chain generic network
described in Sect. 6.3. Stock point 1 with infinite on-hand stock and stock points
20–27, which refer to direct customers, are not controlled in the supply chain. As a
result, the number of stock points with parameters to be optimised in all scenarios
is equal to 33, and the corresponding number of decision variables is 66. Replenishment cycles are defined in days; the minimal replenishment cycle is equal to 7

days or 1 week, and the maximal cycle is equal to 56 days or 8 weeks, which corresponds to one full turn of a planning wheel. Initial stocks are equal to order-up-to levels plus average demand multiplied by cycle delays.

Inputs to the supply chain simulation model specify normally distributed demand for the end-customers and average lead times for processes. For example, an average demand at stock point 42 is equal to 120 product units per period and its standard deviation equal to 12. An average lead time of processes P33–P41 is equal to 2 days while for processes P30–P32 it is equal to 7 days. Inventory holding costs and production costs are defined per product units. Ordering costs are defined per stage. The length of a simulation run is defined by 224 periods or 5,376 hours, which allows modelling of four full turns of the planning wheel.

6.5.1 Optimisation Scenario 1

To solve the problem, the simulation optimisation software add-on SimRunner® is applied, which allows the introduction of the *integer-ratio cyclic policy*. The aggregate objective function is aimed towards maximising a weighted sum of total cost *TC* and fill rate *FR*:

$$\text{Maximise } [Weight_1 * TC + Weight_2 * FR], \qquad (6.9)$$

where weight coefficients $Weight_1$, $Weight_2$ define the importance of objectives (6.1) and (6.2), correspondingly. To set controllable inputs in the simulation model, macros *Cy_i_macro, S_i_macro* are created and used in the optimisation module to define decision variables, and their initial (*Default*) values and decision variable constraints (*Lower Bound, Upper Bound*) are created in the simulation model. For example, *Cy_15_macro* and *S_15_macro* assigned to decision variable of stock point 15 specify the minimal and maximal cycles as 7 and 56 days, and the lower and upper bound of the order-up-to level by 27,020 and 28,030 product units, correspondingly. Fill rate constraints are expressed implicitly in (6.9).

For setting the first optimisation scenario, perform the following steps:

1. Create a new SimRunner® project, then select response statistics, i.e. variables *total_cost* and *fill_rate*, to define objective functions (6.1), (6.2) and specify the total cost to be minimised and a fill rate to be maximised. To define weight coefficients in (6.9), the total cost and fill rate are a priori estimated based on five simulation replications. Then, the relative proportion from received values *TC* and *FR* is used to calculate weight coefficients, i.e. $Weight_1 = 1$; $Weight_2 = TC/FR = 858,180/51.04 \approx 16,814$.

2. Select decision variables, or input factors, from the list of macros, i.e. *Cy_i_macro, S_i_macro*, and for each decision variable define its numeric data type (integer or real) and update, if necessary, the lower and upper bound used to generate feasible solutions in the optimisation procedure. As the number of decision variables in SimRunner® is limited to 60, decision variables for stock points 9, 17 and 41 that have shortest ranges of cycles or order-up-to levels are not optimised.

3. Set optimisation and simulation options. In particular, set the *Optimisation profile* as *Moderate*; and the *Convergence percentage* equal to 0.1. Here, we recommend using the minimal number of experiments as the termination criterion and further stopping the simulation optimisation process manually based on the objective performance plot that shows the convergence of optimisation experiments.

The solutions found (see Fig. 6.10) are ranked and listed according to the values of objective function (6.9). For the best solution that defines optimal replenishment cycles and order-up-to levels, an average total cost is expected to be equal to 904,261 euros and an average fill rate equal to 86.76%. In particular, in this solution cycles for stock points 7 and 10 are equal to 7 days, and the corresponding order-up-to levels are equal to 134,350 and 36,330 product units.

6.5.2   Optimisation Scenario 2

To solve the problem, the simulation optimisation software OptQuest® for ProModel is applied and to define replenishment cycles the integer-ratio cyclic policy is introduced. Here, an optimisation approach is based on the minimisation of the main objective function (6.1) while satisfying customer service requirements applied as a fill rate constraint, i.e.

$$\text{Minimise } E\ (TC) \text{ subject to: } E\ (FR) \geq FR_{min}. \tag{6.10}$$

Let us note, that OptQuest® for ProModel allows introducing not only deterministic constraints on decision variables as in SimRunner®, but also stochastic constraints such as a fill rate constraint defined in (6.10).

For setting this optimisation scenario, execute the following steps:
1. Select decision variables from the list of macros similar to Step 2 in Scenario 1.
2. Define the objective function *total_cost*; and specify constraint *fill_rate*, which is verified after each optimisation experiment is completed.
3. Set optimisation and simulation options similar to those defined in Scenario 1. The solutions with low total cost values that satisfy a fill rate constraint are ranked at the bottom of the output statistics (see Fig. 6.11).

For the best solution that defines optimal replenishment cycles and order-up-to levels, an average total cost is expected to be equal to 869,192 euros and an average fill rate equal to 85.32%. In particular, in this solution cycles for stock points 7 and 10 are equal to 7 days, and the corresponding order-up-to levels are equal to 138,190 and 37,510 product units.

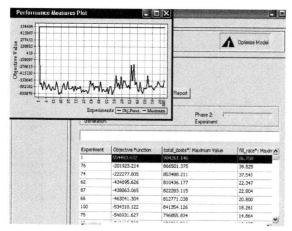

Fig. 6.10 SimRunner® performance graph and output data

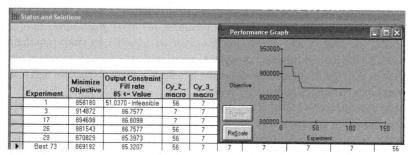

Fig. 6.11 OptQuest performance graph and output statistics

6.5.3 Optimisation Scenario 3

In this scenario, the multi-objective genetic algorithm described in Sect. 6.4.3.1 is used to optimise both replenishment cycles and order-up-to levels, and the power-of-two cyclic policy is applied. The algorithm has the following settings: the maximal number of decision variables is 66, the population size is 40; crossover and mutation probabilities is 0.5 and 0.1, correspondingly; a tournament size is equal to 2 and the number of generations with a stagnant non-domination set that define termination criterion is equal to 3. Decision variable constraints for order-up-to levels are ignored. An initial population automatically includes a starting cyclic solution from a simulation model defined in Sect. 6.4.1.

The optimisation algorithm is developed in MS Excel using ActiveX controls and includes the following worksheets: (a) *User_interface* with three main windows (see Fig. 6.12) to define input data, optimisation and simulation options, (b)

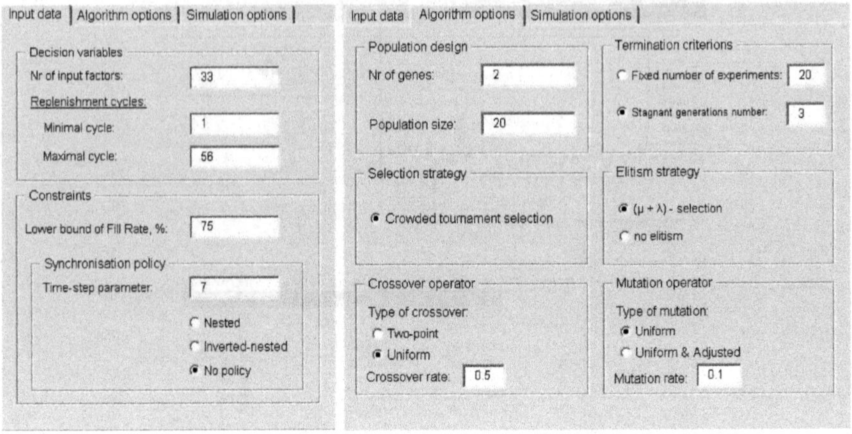

Fig. 6.12 User-interface windows

Network_data to define a supply chain network structure, (c) *Initial_decision_vari-ables* to update initial values of decision variables, (d) *Input_parameters* to update the model parameters and (e) *Output_generations* in order to store populations received during the optimisation process. Control buttons on a user interface allow the user to load the simulation model, calculate the population size, run and manu-ally terminate the optimisation algorithm.

After the simulation model is loaded, Excel worksheets of the optimisation mod-ule that define input parameters and decision variables are updated automatically. After the optimisation algorithm is launched, an initial population is generated and sequentially updated with new generations. Since the GA is stochastic, it is executed five times starting from different initial populations and using different random number seeds in selection and reproduction operations. Then, a composite Pareto-optimal front is generated that includes non-dominated solutions received from optimisations replications.

Examples of initial and final populations in a specific replication mapped in the objective space are given in Fig. 6.13. GA performance graphs in a prog-ress (see Fig. 6.14) show its convergence to solutions with lower total cost and higher fill rate. Finally, the Pareto-optimal set (see Fig. 6.15) received in genera-tions 19–21 contains three solutions with performance average measures: (1) *total cost* $= 787,431$, *fill rate* $= 100.00$; (2) *total cost* $= 766,669$, *fill rate* $= 98.88$; and (3) *total cost* $= 752,300$, *fill rate* $= 93.76$. In particular, in the first solution cycles for stock points 7 and 10 in are equal to 7 and 14 days, and the corresponding order-up-to levels are equal to 170,940 and 55,480 product units.

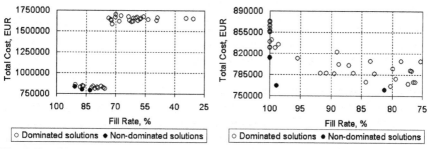

Fig. 6.13 Initial and final population mapped in the objective space

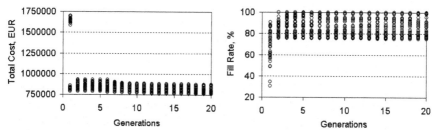

Fig. 6.14 GA performance graphs

	Cy_2_macro	Cy_3_macro	Cy_7_macro	Cy_10_macro	S_2_macro	S_3_macro	S_7_macro	S_10_macro
21	14	28	7	14	187880	28240	170940	55480
21	56	56	28	28	39940	80480	211100	55220
21	14	28	56	28	57950	304080	171100	147340
21	56	56	28	56	56740	87820	265820	294290
21	28	7	28	56	168190	237040	152390	262210
21	7	56	14	28	168200	100690	163390	153720
21	56	56	28	56	220590	89810	265560	73480

Fig. 6.15 A fragment of Pareto-optimal set of solutions in Scenario 3

6.5.4 Optimisation Scenario 4

In this scenario the hybrid algorithm is used in which a genetic algorithm to define optimal cycles of stock points and RSM-based linear search to adjust their order-up-to levels and the power-of-two cyclic policy is applied.

In Phase 1, a genetic algorithm works with 33 decision variables, i.e. replenishment cycles, while initial values of order-up-to levels are calculated analytically. The population size is reduced to 20 individuals. Other GA settings remain the same as in Scenario 3. Five replications are performed for each optimisation experiment. Similar to Scenario 2, the survival probability of infeasible solutions is reduced to exclude them from the population during the evolutionary process.

The initial population includes two non-dominated solutions. However, the diversity of the approximate Pareto-optimal front is increased during the optimisation, and the final population includes seven non-dominated solutions (see Fig. 6.16). The algorithm makes a quick progress during the first five generations, which is typical for the genetic algorithm. Then, there are phases when it hits the local optimum before mutations further improve its performance.

In Phase 2 the approximate Pareto-optimal front that includes seven non-dominated solutions is updated by adjusting order-up-to levels while keeping replenishment cycles found in Phase 1. For each non-dominated solution, RSM-based linear search is performed.

For example, for the second solution in the first iteration the following simulation metamodel is built:

$$TC = 782622 + 43.4_{x_2} + 287_{x_6} + 312_{x_7} + 2.3_{x_8} + 168_{x_{10}} + 141_{x_{11}} + 0.1_{x_{12}} + , \quad (6.11)$$
$$15.4_{x_{13}} + 33.8_{x_{15}} + 15.6_{x_{16}} + 0.2_{x_{17}} + 54_{x_{18}} + 4.7_{x_{19}}$$

where only significant regression coefficients (with p-value < 0.05) are included. Then increments (*Delta*) of input factors S_i and linear search points are calculated according to (6.7), (6.8), and corresponding values of the response *total_cost* are estimated from simulation runs (see Fig. 6.17). A linear search is terminated at the second step when the *fill_rate* objective is decreasing.

Finally, a better solution with the average total cost equal to 781,823 euros and an average fill rate equal to 98.03% is received in the second iteration. In particular, in this solution cycles for stock points 7 and 10 are equal to 56 days, and the corresponding order-up-to levels are equal to 439,360 and 207,780 product units.

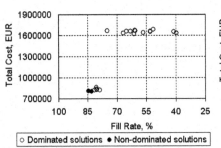 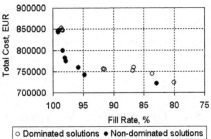

Fig. 6.16 GA solutions in the Phase 1

	S 2	S 3	S 4	S 5	S 6	S 7	S 10	S 42		
Delta	-41	0	0	0	-1784	-2343	-579	0		

	S 2	S 3	S 4	S 5	S 6	S 7	S 10	S 42	Total Cost	Fill Rate
S	58250	37410	129880	59850	387840	468530	214270	37090	782622	98,03
S + delta	57840	37410	129880	59850	370000	445100	208480	37090	781868	98,03
S + 2delta	57430	37410	129880	59850	352160	421670	202690	37090	780433	83,43

Calculate TC and FR

Fig. 6.17 Example of a linear search process

Fig. 6.18 The approximate Pareto-optimal solutions in Phase 2

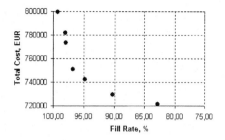

The updated Pareto-optimal front is outlined in Fig. 6.18. There are three non-dominated solutions found in Phase 1 that are improved here by decreasing order-up-to levels in upper echelons. For the other four solutions either the regression metamodel built is not adequate or the total cost could not be decreased without decreasing the fill rate. Let us note that a hybrid algorithm ensures an even spread of solutions along the Pareto front as compared to an isolated GA (see Fig. 6.13), which leads to a broad choice of compromise solutions to be evaluated by a decision maker.

Let us introduce the reference point to define the best values of objective functions based on Pareto-optimal solutions received from the GA and hybrid algorithm: $TC = 721{,}577$ and $FR = 100$. The closest to the reference point solution with performance measures $TC = 721{,}577$ and $FR = 82.88$ belongs to the Pareto-optimal front generated by a hybrid algorithm.

6.6 Conclusions

Optimisation of multi-echelon cyclic plans refers to the class of multi-objective stochastic optimisation problems, which are usually characterised by a large number of decision variables, and multiple and conflicting objectives. An optimisation software add-on that is based on hybridisation of several methods and compatible with discrete-event simulation tools presents broad capabilities to problem solving.

While there is no a single optimal solution for a number of conflicting objectives, the use of algorithms, which give a set of alternative solutions and tackle the response variations generated from the uncertainties in input parameters, is of great practical value. As an example, a multi-objective genetic algorithm based on non-domination concepts is described and applied to cyclic planning and optimisation of a supply chain generic network. A hybrid simulation optimisation algorithm is introduced that integrates a response surface method within the multi-objective genetic algorithm. A hybrid algorithm could outperform an isolated genetic algorithm on the generic network cyclic planning and optimisation as it provides the approximate Pareto front with a wider range of compromise solutions including the solution, which is nearer to the reference point.

6.7   Questions and Assignments

1. What is the difference between formulations of a cyclic planning and optimisation problem in Scenarios 1 and 2?
2. Why are genetic algorithms more preferable among other evolutionary algorithms for optimising multi-echelon cyclic plans?
3. Why can hybridisation of different optimisation techniques lead to a better solution of the cyclic planning and optimisation problem?
4. Change simulation options in Scenarios 1 and 2, for example increase the number of simulation replications or the length of a simulation run. How will the solutions received in each scenario change?
5. Change decision variable constraints by decreasing/increasing their lower/upper bounds by 20%. Run simulation optimisation experiments for all scenarios. How will the solutions and their performance measures received in each scenario change?

Acknowledgement This work was partly supported by the ECLIPS Specific Targeted Research Project of the European Commission 'Extended Collaborative Integrated Life Cycle Supply Chain Planning System' (http://www.eclipsproject.com).

Appendix

To calculate order-up-to levels S_i for each stock point i in each chromosome this sequence of analytical approximate formulas could be used:

$$\mu_{DDLCy_i} = \sum_{k \in succ_i} \mu_{d_{k,i}} * (Cy_i + \mu_{L_{j \in i}}) \,,$$

$$\sigma_{DDLCy_i} = \sqrt{\sum_{k \in succ_i} \sigma_{d_{k,i}}^2 * (Cy_i + \mu_{L_{j \in i}})} \,,$$

$$SS_i = NORMSINV\,(CSL_i) * \sigma_{DDLCy_{k,i}} \,,$$

$$S_i = \sum_{k \in succ_i} \mu_{DDLCy_{k,i}} + SS_i \,,$$

where $\mu_{dk,i}$ is an average demand at stock point k to be satisfied by the stock point i; $\sigma_{dk,i}$ is a standard deviation of a demand at stock point k; μ_{Lj} is an average lead time of process j; μ_{DDLCyi} is an average demand at stock point i during a lead time and replenishment cycle; σ_{DDLCyi} is a standard deviation of a demand at stock point i during a lead time and replenishment cycle; SS_i is a safety stock of a stock point i; Excel function $NORMSINV$ is used to evaluate standard normal cumulative distribution function [13]; CSL_i is a customer service level of stock point i.

References

[1] Merkuryev Y, Merkuryeva G, Desmet B et al (2007) Integrating analytical and simulation techniques in multi-echelon cyclic planning. In: Proceedings first Asian international conference on modelling and simulation (AMS 2007), pp 460–464

[2] Campbell GM, Mabert VA (1991) Cyclical schedules for capacitated lot sizing with dynamic demands. Manag Sci 37:409–427

[3] Merkuryeva G, Napalkova L (2007) Development of simulation-based environment for multi-echelon cyclic planning and optimisation. In: Proceedings 6th EUROSIM congress on modelling and simulation

[4] Abraham A, Jain L, Goldberg R (2005) Evolutionary multiobjective optimisation: theoretical advances and applications. Springer, New York

[5] Azadivar F (1992) A tutorial on simulation optimization. In: Proceedings 1992 winter simulation conference, pp 198–204

[6] Merkuryev Y, Visipkov V (1994) A survey of optimization methods in discrete systems simulation. In: Proceedings first joint conference of International Simulation Society, pp 104–110

[7] Carson M, Maria A (1997) Simulation optimisation: methods and applications. In: Proceedings 1997 winter simulation conference, pp 118–126

[8] Fu MC, Glover FW, April J (2005) Simulation optimisation: a review, new developments and applications. In: Proceedings 2005 winter simulation conference, pp 83–95

[9] Konak A, Coit DW, Smoth AE (2006) Multi-objective optimisation using genetic algorithms: A tutorial. Reliab Eng Syst Saf 91:992–1007

[10] Deb K, Pratap A, Agrawal S et al (2002) A fast elitist non-dominated sorting genetic algorithm for multi-objective optimisation: NSGA-II. IEEE Trans Evol Computat 6:182–197

[11] Goldberg DE (1985) Optimal initial population size for binary-coded genetic algorithms. In: Technical Report TCGA-850001, University of Alabama, USA

[12] Merkuryeva G (2005) Response surface-based simulation meta-modelling methods with applications to optimisation problems. In: Dolgui A et al (eds) Supply chain optimisation product/process design, facility location and flow control. Springer, New York

[13] Chopra S, Meindl P (2001) Supply chain management: strategy, planning and operation. Pearson Education, Upper Saddle River, NJ

Chapter 7
Flexible Manufacturing Systems

M. Àngel Piera Eroles, M. Narciso Farias and R. Buil Giné

Abstract New transport and production resources support high flexibility, resulting in a wide range of options in the planning stage. By increasing flexibility, not only are the number of the decision variables and their domain increased, but the system cause–effect time relationships are as well, which complicates the decision-making activities. In fact, flexibility can lead to benefits but can also lead to idle/oversaturated resources and earliness/tardiness in the final product. The difference between obtaining benefits or losses may depend on the decision-making activity. In this chapter, a discrete-event system modelling methodology to tackle flexibility in present production industries by means of simulation techniques is introduced. The main aspects of the proposed approach using Arena© are applied to remove non-productive operations in a flexible manufacturing system (FMS).

7.1 Introduction

World-wide market competition, high product quality requirements, together with unpredictable demands instead of steady demand are some key factors in this highly competitive market which force the industry to improve its ability to respond rapidly and efficiently to changes in the demand while minimising costs.

Early strategic industrial decisions were oriented towards increasing architectural flexibility (computer numerical control machines, robots, etc.). Thus, new advances in technology allowed production architectures to support flexibility: the ability to handle different product sizes, shapes, weights, paths and volumes with the same equipment. However, several years later, most organisations realised that

Miquel Àngel Piera Eroles, Mercedes Narciso Farias and Roman Buil Giné
Universitat Autònoma de Barcelona, Spain
MiquelAngel.Piera@uab.es, mercedes.narciso@uab.cat, roman.buil@uab.cat

Y. Merkuryev et al. (eds.), *Simulation-Based Case Studies in Logistics*
© Springer 2009

technological flexibility was not synonymous with benefits. Despite that the flexibility to react to market fluctuations can easily be achieved at the operational level by reprogramming production units and transport resources, efficient flexibility can only be achieved by the proper coordination of all the entities (materials and resources), that take part in the production and transport processes.

Unfortunately, technological flexibility has not arrived together with new decision support tools that could allow the industry to take the benefits from this hardware flexibility. Flexibility means choice, so by increasing flexibility, not only are the number of the decision variables and their domain increase, but also the cause–effect time relationships as well, which complicates the decision-making activities.

In fact, flexibility can lead to benefits, but can also lead to idle/oversaturated resources and earliness/tardiness in the final product. The difference between obtaining benefits or losses might depend on the decision-making activity. There are different methodologies that have been used traditionally to respond to planning, scheduling and routing problems; however most of them fall short of offering a proper answer when applied to highly flexible systems.

Simulation models have proved to be useful for examining the performance of different system configurations and/or alternative operating procedures for complex logistic or manufacturing systems. It is widely acknowledged that simulation is a powerful computer-based tool that has enabled decision makers in business and industry to improve operational and organisational efficiency (http://www.dlm-solutions.com/).

Manufacturing simulation models have been widely developed and used for many purposes:
- *Performance prediction*: Checking potential plans and sensitivity
- *Control*: Aiding the selection of desired control rules
- *Insights*: Providing better understanding of the manufacturing system
- *Justifications*: Aiding in selling decisions and supporting viewpoints
- *Optimisation*: Finding the best values for decision variables

However, a word of caution: optimisation should be considered from a more rigorous point of view. When applying simulation techniques to optimise present FMS characterised by non-linear behaviour and a high number of decision variables, subjected to random perturbations, several limitations arise due to its inability to evaluate more than a fraction of the immense range of options available.

In this chapter, the main characteristics of present flexible manufacturing systems will be introduced, along with the main drawbacks of classical simulation approaches. A new modelling approach to quantitatively analyse the cause–effect relationship between manufacturing resources that leads to non-productive operations will be introduced and illustrated.

7.2 Simulation Shortcomings to Improving FMS Performance

A flexible manufacturing system (FMS) [1] is a production system consisting of a set of identical and/or complementary numerically controlled machines that are connected through an automated transportation system. Additionally, each process in an FMS is controlled automatically by a dedicated computer. Under ideal operating conditions, an FMS is capable of processing workpieces of a certain workpiece spectrum in an arbitrary sequence with negligible setup delays between operations. However, setup delays (real operating conditions) can decrease FMS performance results drastically if decision variables, such as processing, handling, storing and transportation, are not well coordinated.

Furthermore, non-value-added operations (such as transporting, storing and inspecting) incorporated in the manufacturing architecture to allow a higher flexibility level are precisely the operations which should be minimised in order to be competitive in time (make to order) and benefits. Intrinsic FMS characteristics that constrain the use of some traditional production planning techniques are:

- *Uncertainty in demand and time production.* FMS production and transport units behave as discrete-event systems.
- *Large number of decision variables.* Note that while flexibility is essential to competitiveness, the number of decision variables that are coordinated and synchronised efficiently is a major drawback.
- *Quick solutions to react to perturbations.* Most optimal planning techniques are CPU-intensive (time consuming) which make them unsuitable to be used for re-scheduling purposes.

To achieve a truly flexible manufacturing system, it is essential to design a control system able to determine the best policy to coordinate both the resources and the flow of products, along with the activities in such a way that non-productive operations (idle times, setting up machines, stocks) can be minimised.

Most commercial discrete-event simulation packages are designed to be used as analysis tools. That is, the system to be studied is modelled, perturbed, parameterised and simulated to predict which changes would cause the disturbances or different parameter configurations in a real system. Figure 7.1 illustrates this approach.

As can be easily understood, the use of simulation models under this experimentation approach to optimising the performance of an FMS is not an efficient approximation, due mainly to the uncountable number of scenarios that should be evaluated, which appear because of the high number of decision variables, necessary for describing the flexibility of the manufacturing architecture, which support the procurement of a final product under a different possible sequence of operations that at the same time can be assigned to different machines.

Furthermore, by developing FMS flowchart simulation models, several shortcomings related to model maintenance and scenario design will appear, because the flow of entities in the system is modelled by the physical connection through the modelling elements (production and transport resources). When the process to be

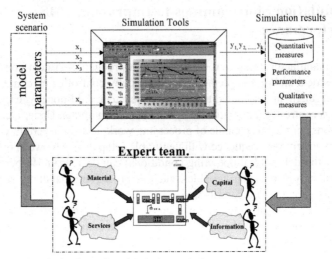

Fig. 7.1 System optimisation using simulation techniques

described can be understood as a specific ordering of work activities, with a beginning, an end and clearly identified inputs and outputs, a model can easily be formalised in a flowchart by mapping elements in the real world into modelling components. However, the efficient evaluation of any routing or scheduling alternative supported by the manufacturing architecture requires a new modelling approach that can easily support the evaluation of different scenarios without forcing changes in the model.

7.3 Managing Simulation Model Complexity

It should be noted that DES (discrete-event simulation) model complexity arises due to a state change as a result of an event that can block, freeze, delay, or disable/enable future enabled/disabled events. A critical barrier to simulation models becoming useful as decision support systems by automatically driving simulations according to inter-medium results is a lack of formalism that could relate to a given system state, and the different alternatives that could be evaluated with the information embedded within the model event-relationships.

Logic constraints between manufacturing resources (processing machines, transport units and local stocks) and production operations, together with their precedence and temporal relationship are some key elements which usually must be formalised in a simulated context as a sequence of events, each one with an associated computer code that upgrades state variables and statistical counters.

Modelling requirements in terms of relationship event specifications requires a knowledge representation technique that considers the stochastic, dynamic and synchronous nature of production systems and allows representing both the structure and the different ways in which a system can behave. A proper representation, analysis and evaluation of all the event-relationships that determine the system behaviour is essential to give a proper answer to industry performance demands [2].

7.3.1 Petri Net Modelling Formalism

Petri nets (PNs) were presented for the first time by Petri (1962) in his doctoral thesis as a formal method for describing computer systems. But the ease with which the PN primitives permitted the description of formerly difficult properties like concurrency, non-determinism, communication and synchronisation, as well as the analysis of these properties, led to the use of Petri nets as true mathematical modelling tools (http://www.informatik.uni-hamburg.de/TGI/PetriNets/).

Their further development was facilitated by the fact that Petri net models easily process synchronisation, asynchronous events, concurrent operations and resource sharing. Petri nets have been successfully used for concurrent and parallel systems and model analysis, communication protocols, performance evaluation and fault-tolerant systems.

A Petri net (see Fig. 7.2) is a directed bipartite graph, together with an initial state called the initial marking. In this graph, there are two kinds of nodes: places (represented by circles) and transitions (represented by rectangles) that are alternatively connected by arcs. An arc can connect either a place to a transition or a transition to a place, but it can never connect two transitions or two places.

Places can contain a non-negative number of tokens, represented graphically as black dots. The number of tokens in a place is the marking of that place, and the array with the number of tokens in every place of the PN (in a certain fixed order) is the marking of the PN. The initial marking indicates the number of tokens corresponding to each place in the initial state. In the PN of Fig. 7.3, left, the marking is M[3, 1, 4, 0, 1].

Petri nets model not only the structure of a system but also its dynamics. This is achieved by changes of state of the PN, which are represented by the evolution of its marking. Thus, the current marking of the net shows the state of the system.

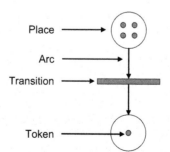

Fig. 7.2 Elements of a Petri net

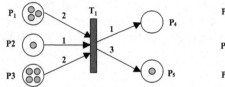 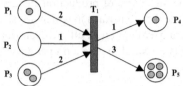

Fig. 7.3 An example of a Petri net

Two special markings are considered: M_0 is the initial marking (initial state of the system) and M_f is the final marking (final or objective state). The change from one state to the next is given by the firing of transitions, which follow the rules below.

7.3.1.1 Rules for the Evolution of Marking

- A place P is an input place of a transition T if there is an arc oriented from P to T. In Fig. 7.3, Places P_1, P_2 and P_3 are input places to the transition.
- A place P is an output place of a transition T if there is an arc oriented from T to P. In Fig. 7.3, Places P_4 and P_5 are output places of the transition.
- A transition is enabled if every input place of that transition got at least as many tokens as the weight of the arc connecting the place to the transition. Thus, the transition of the PN on the left-hand side of Fig. 7.3 is enabled because place P_1 got at least two tokens (weight of the arc connecting P_1 to the transition), P_2 got at least one token and P_3 got at least two tokens.
- An enabled transition is fired if the associated event holds. The firing of a transition implies the removal of a certain number of tokens from every input place and the addition of tokens to every output place. The number of tokens to be removed from the input places corresponds to the weight of the arc connecting the place to the transition. In a similar way, the number of tokens to be added to the output places corresponds to the weight of the arc connecting the transition to the place. Thus, the PN on the right-hand side of Fig. 7.3 represents the new state reached after firing the transition.

7.3.2 Reasons for Using Petri Nets

Given a system, it can be modelled with many different formalisms. The reasons to choose one of the several available specifications are usually related both to the kind of systems to model and to the features of the possible scenarios the model is expected to be used in.

Thus, when developing the model of an FMS to be used as a decision support tool, one of the main aspects to be preserved is to build the most simplified possible representation. The main characteristics of PNs that offer a suitable formalism to describe FMS simulation models are:

- All the events that could appear according to each particular system state can be easily determined (coverability tree).
- All the events that can set off the firing of a particular event can be detected visually.

PNs can be seen as a modelling methodology that supports both characteristics for any type of discrete-event-oriented system, which is essential in improving the performance of complex systems, from the conceptual model that describes all the event relationships to the codification of a simulation model that can support the decision task of optimisation routines at any moment of the evaluation process.

Some other reasons to choose Petri nets as modelling formalism to specify simulation models are:

- Petri nets are a clear, easy to understand and unambiguous modelling formalism. Very little information is needed to synthesise a system, since it includes the concepts of receptivity and sensitivity.
- Given a state, PNs allow us to know the choices to take and the immediate consequences.
- PNs allow the representation of simultaneous evolutions. Thus, parallelism can be modelled and hence, it can be used for the representation of systems with certain industrial decisions to be taken. This property allows the division of the system into different subnets which can represent every set of sequential actions. Therefore, flexibility is included in a Petri net model, since changes can be local to these subnets.
- PNs allow the validation of the right behaviour of the system. The structure and marking of a PN contain information about the system behaviour. This information improves the legibility of the descriptions and the formal validation of certain properties such as detections of deadlocks, traps and failures among others.
- The nets can be generated in a top-down way, by means of continuous refinements. This clearly simplifies PN construction.

7.4   Coloured Petri Net Formalism

Despite all the advantages of PNs as a modelling formalism, there is a drawback to using PNs to describe production, transport, services and logistics systems: a lack of tools to efficiently specify the information flow inherent to any logistics system.

By using colours that allow the representation of entity attributes of commercial simulation software packages, coloured Petri nets (CPNs) allow a higher level of modelling. Other CPN characteristics that enable the use of this formalism to specify FMS are:

- CPNs allow the specification of a system at different abstraction levels, according to the modelling objectives.
- CPNs allow the specification of a complex system by means of bottom-up techniques or more advanced software engineering techniques, such as: an iterative and incremental development process instead of a waterfall cycle, promotion of a component-based architecture.

From the modelling point of view, the main differences between CPN and PN formalism are:

- *Input arc expressions and guards:* used to indicate which type of tokens can be used to fire a transition.
- *Output arc expressions:* used to indicate the system state changes that appear as a result of firing a transition.
- *Colour sets:* determine the types, operations and functions that can be used by the elements of the CPN model. Token colours can be seen as entity attributes of commercial simulation software packages.
- *State vector:* the smallest amount of information needed to predict the events that can appear. The state vector represents the number of tokens in each place, as well as the colours of each token.

Decisions are represented graphically in CPN formalism (i.e. in CPNs) as a place node with different output arcs, each one describing a possible choice. Figure 7.4 shows a node place ($P1$) of a CPN model with several tokens. Arc expressions ($a1$, $a2$ and $a3$) restrict the type of tokens that can be chosen to fire events represented by transitions T_1, T_2 and T_3. Thus, the same CPN model could be used to represent both subsystems ($b.1$ and $b.2$), also shown in Fig. 7.4 just by changing the colours of the tokens in place $P1$ and the arc expressions:

- *Subsystem b.1:* there are three different types of pieces competing to be processed by the same machine. In this case, each transition represents the movement of a certain type of piece to the machine (shared resource).
- *Subsystem b.2:* three different machines are competing to process the raw material stored in $S2$. In this case, each transition represents a different transport operation.

7.4.1 The Coverability Tree

One of the most powerful quantitative analysis tools of PNs and CPNs is the coverability tree [3]. The goal of the coverability tree is to find all the markings which can be reached from a certain initial system state, representing a new system state

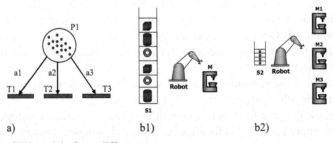

Fig. 7.4a,b CPN model of two different systems

in each tree node and representing a transition firing in each arc. The coverability tree allows [4]:

- All the FMS states (markings) that can be reached starting from certain initial system operating conditions M_0.
- The transition sequence to be fired to drive the system from a certain initial state to a desired end state.

In the first level in the Fig. 7.5, the state vector of a CPN with eight places is represented. In each position of the vector, the tokens and its colours stored in each place node are represented. Given this initial marking, the only enabled events are those represented by transition T_1 and transition T_2. It should be noted that transition T_2 can be fired using three different combinations of tokens (i.e. different entities). Once a transition has been fired, a new state vector is generated.

Thus, a proper implementation of a CPN model in a commercial simulation environment should allow automatic analysis of the whole search space of the system by firing the different sequences of events without requiring any change in the simulation model.

7.5   System Description: a Flexible Manufacturing System

The FMS depicted in Fig. 7.6 (http://tes.uab.cat/FMS) can be deconstructed into three subsystems:

- *Subsystem 1:* Loading/unloading station. It performs the loading of pieces from the initial stock to the pallet and unloading of processed pieces from the pallet to the final stock using a pneumatic manipulator.
- *Subsystem 2:* Manufacturing units, which comprise the following components (see Fig. 7.7):
  - A CNC (computer numerical control) machine with two drills (horizontal and vertical) with different diameters;
  - A local assembly machine with three places for each type of piece;

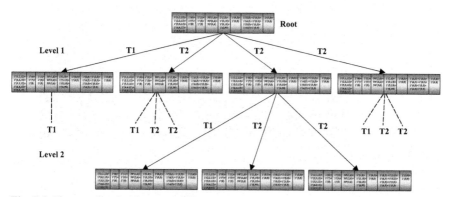

Fig. 7.5 First two levels of a coverability tree

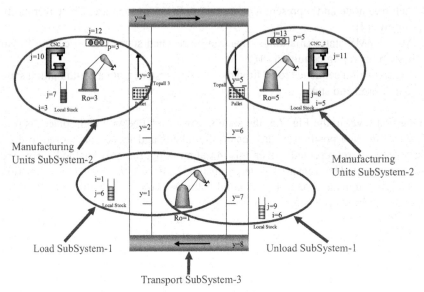

Fig. 7.6 Flexible manufacturing system

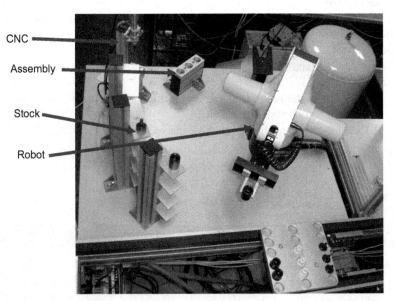

Fig. 7.7 Subsystem No. 2

- A local storage place ('stock') with a capacity of 30 pieces to store processed and unprocessed pieces;
- A robot that transports pieces from one component to the other and to or from the pallet.
- *Subsystem 3*: Transport system that connects the loading/unloading system with manufacturing units. It consists of a conveyor belt with eight positions used to move pallets and six stoppers used to stop and release the pallet at one of the subsystems.

The system is designed to produce different types of pieces according to the different operations that can be configured in the CNC machines.

Table 7.1 summarises the main events of the FMS that should be specified in the simulation model.

Table 7.2 summarises the main attributes of the FMS entities and resources that should be specified in the simulation model.

Table 7.3 summarises the place nodes required to specify the FMS simulation model under the CPN formalism.

Table 7.1 FMS events description

Tr	Description
T_1	Movement of pallet between two consecutive positions of conveyor belt
T_2	Loading pieces from the warehouse of raw material on the pallet
T_3	Transport of a piece from the pallet to the stock of one of the work cells
T_4	Transport of a piece from the pallet to the assembly machine of one of the work cells
T_5	Transport of a piece from the stock of one of the work cells to the pallet located in the work area of the work cell
T_6	Transport of a piece from the stock of one of the work cells to the assembler machine
T_7	Transport of a piece of the stock of one of the work cells to CNC machine
T_8	Conclusion of the operation of processing in CNC machine
T_9	Transport of a piece from the pallet to CNC machine of one of the work cells
T_{10}	Transport of a piece from the pallet to the exit warehouse
T_{11}	Transport of a processed piece from CNC machine to the stock of one of the work cells
T_{12}	Transport of a processed piece from CNC machine to pallet located in the area of work of one of the work cells
T_{13}	Transport of a piece from CNC machine to the assembly machine of the work cell
T_{14}	Transport of a finalised piece from the assembly machine to the stock of the work cell
T_{15}	Transport of a finalised piece from the assembly machine to the pallet located in the work area of the work cell

Table 7.2 FMS colours description

Colour	Definition	Colour description
x	int 1..5	Pallet identifier
y	int 1..8	Pallet position
z	int 1..12	Free capacity on the pallet
j	int 1..13	Piece position in the FMS
k	int 1..7	Piece type or position in the assembly machine
i	int 1,3,5,6	Stock identifier
n	int 0..20	Free capacity in the stock, or busy positions in the assembly machine
p	int 3,5	Available capacity in the assembly machine
b	int 0..1	Free/busy state information
Pa	product x,y,z	Pallet information
Ci	int 1..8	Free position on the conveyor belt
Ro	int 1,3,5	Free/busy robot state
Pe	product k,j	Piece information
St	product i,n	Stock information
Mb	product n,p	Assembly machine busy positions
Mn	product k,p	Assembly machine free positions
Cn	product p,b	CNC machine state: free/busy
Pr	product p,k	CNC processing piece information

Table 7.3 FMS place descriptions

Place	Colour	Description
A	Pa	Pallet resources
B	Ci	Conveyor belt
C	Ro	Robots
D	Pe	Piece information: piece type (1–7), piece position (1–5: in pallet, 6: stock_0, 7: stock_1, 8: stock_2, 9: stock_3, 10: CNC1, 11: CNC2, 12: Assembly Machine_1, 13: Assembly Machine_2)
E	St	Stocks
P	Mb	Assembly machine resources
F	Mn	Assembly machine product characteristics
G	Cn	CNC resources
H	Pr	CNC product characteristics

7.6 CPN Model

By considering that transitions in CPNs can be fired by choosing the appropriate tokens stored in the input places and that the effects of the event are specified by changes in the colours attached to each token, any transition can be translated to Arena$^\copyright$ simulation code (http://www.arenasimulation.com/) in this way:

Tokens stored in place nodes in the CPN model are represented as entities stored in queue blocks in Arena.

Each transition is codified as a sequence of three actions:

1. All the transition pre-conditions are checked by evaluating the attributes (token colours) of the entities stored in the queue blocks that represent the places connected at the input of the transition.
2. Entities that preserve the arc expressions are removed (search and remove blocks) from the queues representing the input places.
3. The attribute values (colours) of the entities (tokens) that have been removed from the queues are updated (assign block) according to the output arc expressions, and are placed again in the queues representing the output places.

To illustrate these codification rules, the transition (T_3) describing the 'Unload a piece from the pallet to a local work cell stock' operation will be illustrated. Figure 7.8 shows the different elements that should be considered to formalise the unloading operation, in which the robot must remove the raw material transported on the pallet to the local stock. It should be noted that transport time for each transport operation can be obtained by means of a deterministic model providing the initial and the final point of the path trajectory. In Fig. 7.8, it has been denoted by two different dotted lines with two different trajectories, each one requiring a particular transport time.

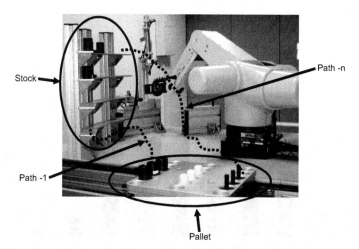

Fig. 7.8 Transport of raw material from pallet to local stock

By considering that the modelling objectives are mainly related to improvement of the FMS performance instead of optimising robot path trajectories, it is possible to use a higher abstraction level in which it is not necessary to formalise the robot dynamics to generate the exact time of each particular transport activity. A random variable is proposed to generate transport times with the same statistic characteristics as the one recorded from the real system.

Once a certain representative number of data sampled from real robot time transport operations has been collected, it is possible to deal with the statistic properties of the sample. Thus, an Erlang (112.57, 0.51) distribution has been found as a function that can be used to model the transport time activity (see Fig. 7.9) independent of the exact initial position in the pallet and final position in the stock. The dark colour is used to represent the theoretical PDF (probability density function) used to describe the transport time, while the other colour is used to represent the histogram of the real data collected.

Figure 7.10 illustrates the CPN model formalising the transport of a piece from a pallet placed in a work cell ($y = 3$ or $y = 5$) to its local stock. In this subsystem, the dynamic to be modelled consists of the movement of the piece from its current position to a new position in the stock ($j = 7$ or $j = 8$). The pre-conditions to allow this movement are:

- The robot of the work cell should be free (tokens in place C).
- The local stock should have at least one free position (tokens in place E).
- A pallet with at least one piece should be placed in one of the work cell areas (tokens in places A and D).

The consequences of a transport operation on the system state variables are:

- The number of empty positions in the local stock is decreased by one unit (token E changes from $1'(y, n)$ to $1'(y, n-1)$).
- The number of empty positions in the pallet is increased by one unit (token A changes from $1'(x, y, z)$ to $1'(x, y, z+1)$).

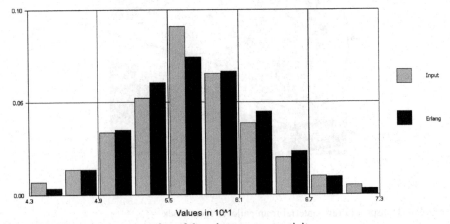

Fig. 7.9 Histogram representation of the robot transport activity

- The new location of the piece is the stock (token D changes from $1'(k, x)$ to $1'(k, 7)$ or $1'(k, 8)$).

Figure 7.11 illustrates the Arena code for transition T_3.
The functionality of each block is:

- *Decide 4:* checks that the robot $ro = 1$ is free. In case the robot is busy or working, transition T_3 cannot be fired, and the entity is sent to a dispose block.

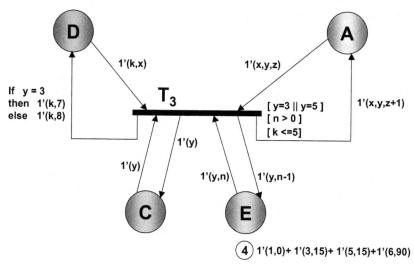

Fig. 7.10 CPN model of transition T_3

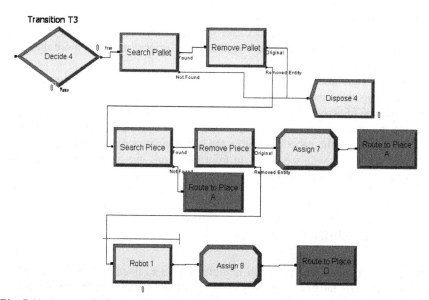

Fig. 7.11 Arena code for transitions T_2 and T_{10}

- *Search pallet:* searches in the entities stored in queue place *A* (which represents the information associated with each pallet), if there is a pallet in position 3 or 5 with at least one piece in the pallet. If a pallet is found, block 'Remove pallet' is removed from queue place *A*, the entity representing the pallet placed at position 1 on the conveyor belt.
- *Search piece:* searches in the entities stored in queue place *D* (which represents the information associated with each piece in the FMS), if there is a piece in the raw material stock (a token with colour $j = 6$). If a piece is found, block 'Remove piece' is removed from queue place *D*, the entity representing the piece placed at the raw material stock. It should be noted that if we were interested in loading a certain type of piece, an extra condition in the 'Search piece' block could be added indicating the type of piece.
- *Assign 7:* the information related to the number of free positions in the pallet is updated according to the CPN model (attribute *y* is decreased). Once the attribute of the entity representing the pallet has been updated, the entity is sent to place *A*.
- *Robot 1:* represents the resource that will perform the load operation, consuming a certain amount of time.
- *Assign 8:* the information related to the position of the piece in the FMS is updated according to the CPN model (attribute *k* is changed from 6 to the value of the pallet identifier). Once the attribute of the entity representing the piece has been updated, the entity is sent to place *D*.

7.7 Results

The simulation model in Arena can be set up quite easily just by adding the Arena code of each transition specified in CPNs. A decide block should contain the scheduling event policy in which the events should be prioritised when more than one event could be fired.

Different simulations have been performed just by changing the priority of the different transitions modelled in CPNs in order to determine which would be the best sequence of activities that would drive the system from its initial state (workload) to the desired state. Figure 7.12 shows the scheduling results obtained when trying to optimise the robot's resource for a particular workload. To evaluate the robustness of the scheduling policy in front variations due to random aspects, different simulations should be performed.

7.8 Conclusions

The flowchart modelling approach is inefficient in evaluating a high number of scenarios because the flow of entities in the system is modelled by the physical con-

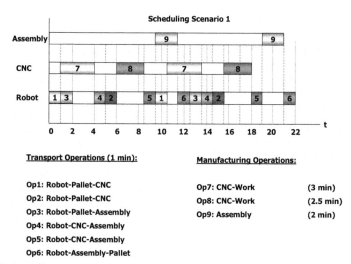

Fig. 7.12 Gant diagram

nection through the modelling elements (production and transport resources). Under the flowchart modelling approach, the evaluation of a new transport or scheduling alternative requires a new simulation model to describe each new scenario.

The main advantage of using the proposed modelling approach is that the same simulation model can be used to evaluate the different scenarios in which the FMS could be managed. It should be noted that the different scheduling and/or planning policies are codified in the Arena search blocks placed in each transition code. Thus, just by changing a variable value (the type of entity to be processed), it should be possible to evaluate all the scenarios, without changing the simulation model. This approach can be achieved automatically if the variables used in the search blocks are updated using an external file that specifies the scenario to be evaluated.

The robustness of the scheduling policies obtained by the proposed approach should be evaluated in a posterior phase by means of a multi-run simulation in which the variability's effects on the stochastic time models would be properly represented.

7.9 Questions

1. The simulation model in Arena looks like a flowchart modelling approach. What are the differences introduced by the CPN formalisation?
2. Why do you think that model maintenance is easier using the proposed approach?
3. What are the main aspects of industrial flexibility that constrain the use of simulation techniques to deal with optimal scheduling policies?

References

[1] Zhou MC, Venkatesh K (1998) Modelling, simulation, and control of flexible manufacturing systems. World Scientific, Singapore

[2] Piera MA, Narciso M, Guasch T et al (2004) Optimisation of logistic and manufacturing systems through simulation: a colored Petri net-based methodology. Simul: Trans Soc Model Simul Int 121–130

[3] Proth JM, Xie X (1996) Petri nets. A tool for design and management of manufacturing systems. John Wiley, New York

[4] Jensen K (1997) Coloured Petri nets: basic concepts, analysis methods and practical use, vols 1–3. Springer-Verlag, Berlin

Chapter 8
Fresh-Food Supply Chain

A. Bruzzone, M. Massei and E. Bocca

Abstract Modelling and simulation (M&S) is a critical technology when applied to complex logistics; it is evident that food and especially the fresh-food supply chain [1] represents a very interesting application area, considering all the inter-related constraints and variables: time-to-market, traceability, transport/storage conditions, handling, production/process control, demand variability, seasonal behaviours, etc. In fact, food represents a strategic sector; in order to increase margins on specific products such as red and white meat, fresh fish, fruits, vegetables, frozen foods and dairy products, an effective management of the logistics operation costs and food preparation is needed in order to develop new solutions for these special supply chains. This approach requires development of logistics models in order to achieve different results such as faster distribution processes, rapid response with cost reduction, and increase in good useful lifetimes. The chapter proposes a fresh-food supply chain model devoted to support logistics network re-engineering as well as operation management.

Agostino Bruzzone
University of Genoa, Italy
agostino@itim.unige.it

Marina Massei
Liophant Simulation, Italy
massei@itim.unige.it

Enrico Bocca
MAST Srl, Italy
simulation@mastsrl.eu

Y. Merkuryev et al. (eds.), *Simulation-Based Case Studies in Logistics*
© Springer 2009

8.1 Introduction

Among the most relevant critical aspects affecting fresh-food products, it is easy to identify several specific aspects such as:
- Perishability, which implies the need for very rapid logistics processes.
- The very high profile needed from an organoleptic quality and freshness point of view, which constitutes hard constraints for cost reductions.
- Traceability, required among the goods to be distributed as well as controls for guaranteeing safety and also for securing the supply chain.
- The special processes required for preparing food along the supply chain: for instance, slaughtering, meat cutting, packaging or modified atmosphere packaging (MAP).
- Strong seasonal behaviour of demand and production, which introduces a continuous evolution in the product mix as well as the necessity to organise a robust and flexible logistics network [2].
- Difficulty in creating an efficient and optimised platform due the interaction among many logistics flows (many supplier deliveries to be divided and mixed for shipping to many shops).
- Presence of direct distribution flows from producer to final consumer (stores).

Considering these factors the problem of very rapid logistics is really challenging, especially considering concurrent needs for guaranteeing a high level of service to the customers, while maintaining an efficient system of safety/security controls on the products, and reducing costs.

For instance, in the case of retail companies serving stores [3] the logistics solution needs to satisfy many constraints looking to optimise multiple target functions; in fact these realities operate often on diverse scenarios: they need to serve different kinds of stores characterised by size, customer profiles and cultural gastronomic backgrounds related to the different geographical locations (urban areas vs. rural, coastal vs. inland zones). This complexity often requires developing tools for decision support and logistics management based on M&S. In fact in this context it is critical to define a strategic logistic view for identify the right solutions considering all possible trade-offs in term of time control, quality levels, customer satisfaction and economic profitability. These decision support systems are not only devoted to supporting supply chain design or re-engineering, but also to defining/maintaining the performance reference baselines and metrics to be used in overall process control [4].

8.1.1 Fresh-Goods Processing

From production sites to final customers, the flow of fresh goods is processed and moved along the different phases of the supply chain, while in the opposite direction information flows are used for driving the planning and distribution.

An effective way to increase the efficiency of the whole system is related to the improvement of the integration of these two flows for the overall optimisation of the supply chain [5]; this result is often achieved by electing a centralised solution (physical or virtual) in terms of operations, goods stocking and company structure. In particular, using M&S, it is possible to develop and evaluate smart alternative logistics solutions [6] devoted to taking advantage of synergies among different goods, sale channels, purchasing offices, suppliers, geographic areas, etc. Fresh-food deliveries from sources (farms, breeders, sea, etc.) to the final customers (typically hypermarkets/supermarkets, small stores, etc.) introduce many different options in terms of different activities and process locations: for instance in fresh-meat logistics, the slaughtering can be performed by the producer, or carried out on the distribution platforms; these two alternatives involve different costs/benefits in terms of quality control, costs and overall process efficiency that are very hard to evaluate without a detailed simulation that is able to consider the different goods (beef, pork, etc.) and cuts (shoulder, tenderloin, etc.) [7].

In fresh-food supply chains it is necessary to develop solutions that can guarantee an effective control system for integrating into company enterprise resource planning (ERP) all the logistics flows and fresh-food platforms [8]. This logistics network includes suppliers, transports, logistics and operative platforms, warehouses and stores. Different alternatives, including cross-docking, multi-drop and shuttle services need to be evaluated in order to identify best supply chain configuration. From this point of view M&S is the ideal methodology for defining solutions to such problems and for analysing different scenarios. The authors propose a case study where M&S was successfully used to analyse fresh-food logistics for meat and fish goods [9].

8.1.2 Logistics Solutions

A common solution for logistics improvements is storage reduction; in fresh food this is often integrated with the creation of cross-docking processes (not only logistics, but even operations on the food products). The cross-docking approach makes possible the transfer of incoming shipments directly to outgoing trailers, without storing them in between, and eventually to process the food on the way (i.e. special mix production, or packaging) in order to create scale economy advantages. In retail, cross-docking processes often connect suppliers to stores, defining the distribution of the goods based on pre-defined criteria in order to face contingencies (accidents, quality screening, etc.) [10]; in this case the logic flow is defined as: arrival of goods from suppliers, assignment to the store, goods processing, goods delivery to stores. Often all these steps do not require any stock on the platform, and minimise handling operations; sometimes some storage is required but the crossing time of the platform is very commonly reduced to a few days. This approach increases the useful lifetime of products and keeps under centralised control quality and safety, on the basis of acceptance control. Usually these platforms are requested to serve concurrently different sale channels (supermarkets, hypermarkets, small stores, etc.) and geographic areas; in this case, the incoming supplier flows need to be composed of a good mix able to satisfy the different requirements in terms of store format: packaging quantities, quality, marketing target, geographic region, etc. In addition to this aspect it is necessary to consider the opportunities in shifting processes: for instance in meat distribution the processes related to slaughtering, preparation and portioning as well as packaging, can be shifted from supplier to the final store or on the platform, as summarised in Table 8.1, which presents possible advantages (Adv) and disadvantages (Dis) for each alternative.

The reduction of costs in introducing a platform along the supply chain vs. direct shipping to stores is mostly based on the discounts that suppliers offer to retailers due to the improvement in their logistics (delivery to a single point vs. a large network of destinations) and in their commercial network.

In addition for retailers, this solution guarantees a more effective control on quality and delivery times with opportunities for additional saving in centralised goods processes by moving from mere logistics platforms to more flexible platforms including operations to be completed on-line without requesting real warehousing (eventually very short-term storage).

Obviously these platforms introduce fixed costs related to the infrastructure and variable costs due to the operations, so case by case it is fundamental to complete a detailed trade-off analysis by developing and executing proper models.

In general the customer satisfaction could increase due to the fact that products have a longer lifetime since they do not incur days in storage, and delivery is punctual and based on the original store request, reducing any risk of stock-out; therefore centralisation for goods of this kind moves to a more 'industrial' process. However, this could sometimes result in a negative impact for the consumers [11]; for instance the fact that the butcher is no longer active in the store could result in the perception of a less fresh product or in losing expert advice at the store counter on choosing and preparing meat.

Table 8.1 Red-meat processes

	Supplier		Platform		Store	
	Adv	Dis	Adv	Dis	Adv	Dis
Slaughtering	Easy to manage	Low Control	High control	Medium cost Transportation	–	Too expensive
Portioning	Flexible response	High cost Impact on consumers	Low cost	Impact on consumers	Good impact on consumers	High cost
Packaging	Flexible Response	High cost Impact on consumers	Long duration Low cost	Impact on consumers	Longer duration	Impact on consumers

These aspects deal with marketing and commercial considerations; therefore the authors currently just focus on quantitative logistics and operative aspects, while these non-quantifiable issues need to be evaluated by experts with respect to the obtained figures related to cost saving and process control improvements. So the locations of fresh-food processes along the supply chain need to be defined among the possible alternatives based on strategic issues and related costs [12]. M&S requires consideration of all scenarios and to computation of the performance and costs for the different solutions; for instance if all the food processing is carried out by suppliers, the retail operator just takes care of sale activities, so direct delivery is potentially more effective and a 'mere logistics cross-docking point' is probably the best alternative to be considered for saving on transportation. Vice versa, if the goal is the reduction of costs related to goods processes the platform will have probably to take care also of portioning and packaging at least. The authors developed a case study for the fresh-meat supply chain of a major retailer operating in northern Italy; in this case a special simulator was developed implementing all the above-described criteria and alternative solutions (http://st.itim.unige.it/projects/; http://www.liophant.org) and in the following it is proposed in some further detail.

8.2 Meat Distribution Simulator

The case proposed is related to a regional supply chain for fresh meat covering an area populated by about 15 million inhabitants; the goods considered cover all the kinds of red and white meat (beef, veal, pork, horse, lamb, poultry, etc.), while the geographic area includes very different zones (a large town with over one million inhabitants, rural areas, tourist Riviera and Alps); these differences correspond to different demand profiles evolving over the year in different ways introducing

Fig. 8.1 Meat distribution simulator reproducing all communication, production, transportation and logistics processes

additional complexity in the supply chain management; due to these factors it was decided to develop a simulator able to take care of all the different factors; Fig. 8.1 shows a graphical representation of the user interface.

In fact the model needs to be able to consider even multiple daily dispatches of goods since commercial experts consider this an opportunity for new promotions.

In the following is proposed the model developed for optimising [13] these aspects for food products; in fact the logic procedure is defined as follows: requests need to be submitted to central management in time in order to be distributed to logistics operators, checked and processed. Calendars for goods delivery to the stores are estimated based on supplier lead time and delivery calendar. The proposed modelling approach considers also an additional order calendar for such fresh food, in addition to the previous one that combines platform operative times with store requirements. In order to keep under control the quantities ordered, platform operators have to use dynamic support lists, taking into consideration fundamental 'key words' such as product type, quantities and sales points. Such parameters are updated by the ERP system where users can change distribution quantities/products (equivalent products from different suppliers) or add information to avoid common operative mistakes, often related to the product submittal calendar, and ordered quantities [14].

The conceptual ordering model includes direct orders to suppliers: in this case, a dynamic list system has been designed to support transmission of orders from stores through the central purchasing office to the suppliers, keeping up-to-demand evolution. Such a system proposes quantities of goods to be ordered, considering: original store requests, available stocks, backlog orders, late orders from the store, assignment percentages, market price, and special discounts from suppliers.

Store managers can change such proposals based on their preferences, therefore in order to keep the process under control, the logistics network needs to define rules for assignments and changes (maximum change to an original order from a store); these policies strictly depend on the network structure (stores and platform belonging to the same company) and strategies (centralised control versus maximum store autonomy/responsibility) [15]. It is critical to define for the real case a set of algorithms for correlating these factors and generating the proposal; these algorithms operate according to the product mix characteristics and prices and they determine the proposed ordered quantity for each case; for instance the issues related to multiple suppliers of the same item are implemented by the strategies of Table 8.2.

Table 8.2 Single and multiple supplier management strategies

Multiple supplier	If several suppliers are associated with the same product and none of them is identified as a 'regular supplier', the algorithm assigns a share of the request from the customers in relation to a pre-defined percentage assigned to that supplier. The residues of the request are assigned to the suppliers for which the proposed quantity is not rounded off to a pallet multiple. Any residues are assigned to the suppliers with a lower purchase price.
Single supplier	If a supplier is identified as a 'regular supplier', the algorithm assigns the quantity requested by the sales points to this one.

The overall process can be summarised as the following:

1. The order is estimated by demand-forecasting models, so the orders are created by an ERP transaction that generates reports for all the items expected to be demanded by stores and the relative suppliers.
2. The quantity to be ordered is estimated in order to cover the requests for the good lead time by predictive algorithms based on a weighted mean of consumption computed over homologous days of the last five weeks and consequently reallocated over the suppliers based on specific shares and delivery calendars.
3. The sharing of the items among suppliers is determined by an 'accumulation mode' that considers the expected ordered quantity based on previously dispatched orders.
4. The orders are automatically submitted, via fax, electronic data interchange (EDI) or e-mail, but it is also possible to manage additional channels.

Since availability levels can increase or decrease based on many different factors, suppliers are often unable to fully satisfy the requests; for this reason a set of functions to manage this kind of 'emergency' has been developed (extra orders to local suppliers to be delivered directly to stores). Every day, the suppliers have to notify stores about unavailability of specific goods by noon, so that the proper corrective actions can be taken: redirection of part of the ordered goods to another supplier, or redirection of the request to another similar reference so that the sales points can avoid stock-outs. If an order is shifted to an unplanned supplier, the quantities delivered have to be used for covering the demands of stores that are included in the last dispatch mission from the platform; this policy is motivated by the fact that such redirected quantities are expected to arrive at the platform among the last goods of the day (being late orders). If it is impossible to compensate the supplier stock-out, these unavailable quantities need to be shared over the store network by specific algorithms; the authors developed a special module that includes the possibility to manage such problems both manually and by automatic redistribution functions. Automatic redistribution functions are based on two basic different algorithms that can be combined:

- *Card rule*: this uniformly redistributes the quantities over the sales points in the same way that a deck of cards is distributed among a group of players; as a consequence of this approach the requests of the smaller sales points are usually fully satisfied, to the detriment of the larger ones.
- *Proportional algorithm with minimum threshold*: this distributes products in proportion to the original requests, guaranteeing to satisfy at least a predetermined threshold level. Applying such an algorithm favours the satisfaction of sales points requesting a huge quantity of goods.

The authors decided to proceed with a proportional algorithm with minimum threshold; this proportional algorithm was designed in order to operate through a sequence of three actions that are described in the next three algorithms.

8.2.1 Redistribution Algorithms

In the following is proposed an approach for defining how to share over-/under-delivery among stores in order to manage contingency in the supply chain. These algorithms are related to each store (i) and each item order (j) in order to define request thresholds and determination of residuals; the use of the simulation allowed us to tune and validate these approaches:

- If $PoS_{i,j}^{Request} < Threshold_j => PoS_{i,j}^{RequestThreshold} = PoS_{i,j}^{Request}$,

- If $PoS_{i,j}^{Request} >= Threshold_j => PoS_{i,j}^{RequestThreshold} = Threshold_j$,

- $PoS_{i,j}^{RequestResidue} = PoS_{i,j}^{Request} - PoS_{i,j}^{RequestThreshold}$,

- $TotalRequestThreshold_j = \sum_{i=1}^{k} PoS_{i,j}^{RequestThreshold}$,

- $TotalRequestResidue_j = \sum_{i=1}^{k} PoS_{i,j}^{RequestResidue}$,

where $PoS_{i,j}$ is a good request for the jth item from the ith store; $Threshold$ is the minimum threshold for the jth item order; k is the number of the store.

The indicator calculation is proposed in the following for each item (j):

- $ThresholdPercentage_j = \dfrac{Min(ThresholdPercentage_j; Available_j)}{TotalRequestThreshold_j}$,

- $ResiduePercentage_j = \dfrac{Available_j - Min(TotalRequestThreshold_j; Available_j)}{TotalRequestResidue_j}$,

where $Available_j$ is the quantity provided by the suppliers of the jth item.

If residue $Percentage_j$ is greater than one, this field has to be forced to one and the extra quantities delivered remain in stock in the warehouse.

The assignment and the residue are defined as following:

- $PoS_{i,j}^{ThresholdAssignment} = int(PoS_i^{RequestThreshold} * ThresholdPercentage_j)$,

- $TAT_j = \sum_{i=1}^{k} PoS_{i,j}^{ThresholdAssignment}$,

- $PoS_{i,j}^{Residue\ Assignment} = int(PoS_i^{Request\ Residue} * ResiduePercentage_j)$,

- $TAR_j = \sum_{i=1}^{k} PoS_{i,j}^{ResidueAssignment}$,

where TAT_j is total assigned threshold for the jth item; TAR_j is total assigned residue to the jth item; $int(z)$ is the integer part of z.

Subsequent to the previous assignment of quantities, eventual residues to be assigned and are attributed as follows:

- $SDT_i = Min\ (TotalRequestThereshold_j; Available_j) - TAT_j$,

- $SDR_i = Available_j - Min\ (Total\ RequestThereshold_j; Available_j) - TAR_j$,

where SDT_j denotes still-to-distribute *Threshold_j*; SDR_j denotes still-to-distribute *Residue_j*.

Designing the delivery procedure in order to complete the preparation from stock processes requires us to define a proper model [16]; it is necessary to take into consideration also the case that one lot remains in storage at the end of the last distribution. Such a lot has to be distributed to the sales points due to the heavy constraints in the expiration date for fresh foods.

In order to avoid the assignment of 'old' lots systematically to the same stores, an ad hoc algorithm has to be developed. Such a distribution algorithm is based on saturation of the store request, starting from the sales points not yet served.

8.3 Fresh Fish: Definition of Delivery Processes

A very critical step in fresh-food logistics is the definition of delivery processes; here is presented another case related to the distribution of fresh fish to supermarkets/hypermarkets over a large area composed of three regions; also in this case the design of the supply chain structure is critical to guarantee cost saving and quality improvements.

In fresh-food logistics for big supermarket chains it is possible to use a centralised distribution platform to receive all goods from all suppliers and, after operations or cross-docking, to serve all final stores. In this case the delivery process is composed of two main phases presented in Fig. 8.2: a component called the first ring, including the section supplier/platform, and a second ring related to the connection platform/customer. In this case the first question is how many platforms to set up and where to locate them.

SUPPLY CHAIN

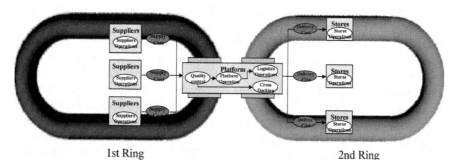

1st Ring   2nd Ring

Fig. 8.2 Example of supply chain model based on two rings for fresh fish

Therefore the main supply chain improvements are based on a smart management of synergies between the two rings: for instance reorganising supplier deliveries to the platforms that are more convenient is a way to reduce prices by improving saturation indexes and to create opportunities for demand growth by proposing to the consumer more rationalisation of product mix. Obviously an easy way to create additional savings (involving logistics and commercial issues) is related to the rationalisation of suppliers, decreasing their number and increasing their individual volumes. At the same time a well-organised second-ring optimised distribution management results in defining a set of missions (multi-dropping platform to stores), optimised in terms of cost and services; however, this is not an easy goal due to constraints related to the store distribution; in fact, transportation and docking of fresh food are subject to legal rules to preserve the quality of the products. Thinking, for example, about fresh fish, the European Union and other organisations set out a list of rules, in particular about thermal regime, while fresh-fish deliveries in a capillary scenario composed of super-/hypermarkets require frequent load/unload stops having heavy impact on temperature evolution in the refrigerated trucks; this behaviour could result (if the route is composed of short distances between many destinations) in exceeding admissible temperature ranges; this event introduces the probability of losing quality corresponding to economic and financial penalties (in addition to commercial and image impacts). In fact temperature control is very significant in operations flow because missions are classified typically into two different phases: one is where temperature is lowered through uninterrupted use of refrigeration systems on the trucks (during driving, where cell doors are closed and the only airflow is from cooler to products), and the other is the collection of activities that determines raising of temperature, such as loading/unloading of goods. It is therefore important to combine operations to in a complementary way that can provide the maintenance of the temperature in an acceptable range. This is possible if there is sufficient time, thus distance, between consecutive stops.

Other constraints affect traditional transportation issues, such as working time limits for drivers (including stops), and problems regarding access timetables for the stores; in fact deliveries are usually forced into pre-defined timeframes consistent with personnel availability and urban framework constraints. Elapsed time at each stop depends on load/unload operations and changes in relation to the kind of good and package (pallets or boxes) and store infrastructure (possibility to use loading/unloading bays or need to use truck elevator/special transpallets).

For these reasons even simple mission definition in compliance with described constraints is a not an easy job; obviously the optimisation is very challenging considering the necessity to proceed towards following three major steps: clustering stores (deciding what stores to serve with a single mission of a truck in a day), sequencing stops (definition of the sequence of service to the stores during a mission) and scheduling missions (defining on which day of the week and at what time each mission starts). In fact, specific technology improvements in logistics and transportation help subjects to relax some constraints by investing in more efficient infrastructure/carriers; for instance upgrading of loading/unloading bays reduces service time in stores and improves mission efficiency, in this way it is possible to save time, reducing costs and temperature risks. Other examples are equipping receiving gates with ramps to give fast and direct access to trucks, and using pallets to better stock goods, limiting heat growth; at the same time it is possible to invest in insulating of dock gates, reducing swapping with external air.

The logistics of fresh food require simultaneous delivery of items with different characteristics, and they need specific conditions to be stocked and transported. In fact, for example, different species of fish require different thermal values (fresh fish is transported at 0°C while frozen fish is transported at −18°C to −20°C), and technology can help delivery with a single mission using particular trucks, equipped with multi-temperature cells. This solution provides cost savings by reducing the number of carriers, but incurs increasing difficulty and time in loading/unloading operations, limiting the possibility to use pallets (if the deliveries are organised per store is impossible to set a pallet with items needing different temperatures), and with limited accessibility to goods (multi-temperature cells are not completely accessible from the back of the cell). All these alternatives require trade-off analysis where simulation is the proper approach to providing quantitative evaluations.

A good approach to finding a solution is to use simulation-based models able to reproduce all peculiarities and providing the decision makers with reliable data about different alternatives; the expectation performance is used in reality to provide a reference baseline for logistics operators; these values during the implementation in reality allow the supply chain to be kept under control and optimal operations to be defined, i.e. store resequencing over contingencies as well as estimation of costs, risks, quality levels and all significant indexes (frequency or on-time arrival levels, vector saturation, temperature course, etc.) [17]. A logistic solution based on independent first- and/or second-ring optimisation represents a local suboptimal solution, often with hard problems in real implementation: in fact it is necessary to define the logistics solution and to work to create a real combined optimisation that generates synergies between the two rings. A problem in combining the two rings is related to goods arrivals and departures not in synchronisation that, considering the impossibility to maintain stocks due to the fast-perishing nature of fresh fish, forces second-ring deliveries (to stores) to be strictly conditioned by arrival times from suppliers while they need at the same time to respect store access time gates.

Another common problem deriving from searching to optimise the first and second ring independently arises through incompatibility between demand and offer platform/store: to increase advantages relating to the second ring probably every platform will serve only a cluster of customers, therefore to improve first-ring supplies it is convenient to have only one distribution centre receiving from each supplier, optimising vector saturation. The two aspects are not consistent with a simple solution; however, it is possible to define a configuration using a shuttle service as an 'internal transport service' platform to platform to compensate demand/arrivals; this service needs to be properly evaluated in terms of cost and service, which changes over the week and the year and could generate unacceptable solutions.

In fact it is critical to create a model not only about direct activities for logistics services (trucks, drivers, loading/unloading) but even for infrastructures and indirect costs that are required (goods-handling platform infrastructure, extra transportation for platform synchronising problems, etc.). The model developed considers risks and extra cost to compensate contingencies (extra demand that requires additional transportation several times a year).

The stochastic model created is based on simulators devoted to reducing risk of incurring undesired situations and providing good control of cost/performance. In fact the logistics of fresh food includes a high cost related to commercial and logistic issues (discount from suppliers) to be optimised looking for an efficient focus on different payoffs.

The simulator allows costs/benefits to be quantified for each factor; for instance it is possible to define the saving in reducing the risk of exceeding the temperature threshold versus costs due to the reduction of truck saturation level in the second-ring phase.

The simulation model reproduces the mission planning defined by user (or optimised by some smart planning system) and estimates all the costs related to temperature risk as well as to all the other issues.

The model requires defining criteria for each component, for instance the cost of exceeding the temperature threshold.

In fact it is important to fix the criteria for each critical event and its consequences; for example it is possible to consider the risk of a late delivery to a store during a mission. As the first hypothesis it is possible to consider two basic eventualities: necessity to cancel the delivery or extra payment for finalising this delivery. In fresh fish, operating without stocking, the cancellation of the delivery corresponds to a financial loss equivalent to the value of the goods, to be further increased by the necessity to dispose of the goods rejected. Vice versa, to complete the late delivery the company has to sustain extra costs, for example to pay the driver and the personnel to receive items in store as extra time. A simple algorithm to define the value of this eventuality is

$$x_l = p_r * \left\{ p_l * v_l + p_p * \left[t_l \left(c_d + c_t + \sum_{i=1}^{n} c_{s_i} \right) \right] \right\}, \qquad (8.1)$$

where:

- x_l = late time extra cost
- p_r = late time risk probability
- p_l = delivery loss probability
- v_l = goods lost value
- p_p = pay to delivery probability
- t_l = expected late time
- c_d = driver cost per hour

- $\sum_{i=1}^{n} c_{s_i}$ = structure additional costs per hour

- c_t = truck cost per hour

Estimating these values for every kind of risk allows use of the simulation model as support for decision makers in order to optimise logistics strategies.

8.3.1 MARLIN Simulator

In the proposed case, a special architecture integrating different simulators was developed for combining the different supply chain phases and a smart planning system able to identify optimal solutions by applying artificial intelligence techniques (in this case genetic algorithms). In fact, this research corresponded to a real project: Models for Advanced Reorganisation in Logistics of Ichthyic Nourishment (MARLIN). MARLIN [18] was devoted to facing the challenge of reorganisation in a complex logistics network for fresh-food distribution by applying innovative simulation models; the project was motivated by the necessity to test the potential of these techniques in a realistic case study: MARLIN is presented here with reference to a fresh-fish supply chain. This context is based on real estimations on this specific supply chain provided by experts; the logistics processes of fresh foods are based on the two rings mentioned above: the first one interconnects suppliers and fishing companies to the central management system (composed of the different logistics platforms located in diverse areas and the centralised purchasing office); vice versa, the second ring connects the logistics platforms with the stores that are spread out in the area under analysis, which is divided into three different geographical zones (two inland urban areas and one coastal zone). The store network is organised based on two different kinds of sales channels: supermarkets (up to 2,500 m²) and hypermarkets (over 2,500 m²). The fish demand in this context is very different due to the variability of the customer profiles in the two sales channels and due to the gastronomic and cultural background within the different geographic zones. It was therefore necessary to define classes for the different store zones, so it is critical to consider in the models the different quality perceptions and consumer needs for each class; the final distribution configuration, for the proposed case, is composed of about 120 stores with detailed data for each one (access constraints, delivery time gate, loading/unloading facilities, dynamic demand composition, preferences and intensity stochastic definition, etc.) [19]. The results are very variable also among the different stores of each zone (Z1, Z2, Z3) and of each sales channel (hyper-/supermarkets) as shown in the graphs of Figs 8.3 and 8.4.

The following supply chain solutions were considered as different scenarios:

- Scenario 1: Three different independent logistics platforms located in each zone and serving local stores with supplies from all providers
- Scenario 2: One single logistics platform covering all zones
- Scenario 3: Three platforms interconnected by shuttle services

In this case the medium–short-term goal is to move all the fresh-food logistics flows through platforms, without any remaining direct delivery to stores, in order to guarantee full control of the process; so all the direct deliveries need to be redirected through the platforms. This introduces additional handling costs, which are potentially compensated by commercial discounts (centralised order) and logistics discounts (delivery to a central point); obviously it is necessary to simulate the scenarios to identify the most promising solution in this specific context and with current hypotheses in term of costs and fees. At the beginning of this project each area

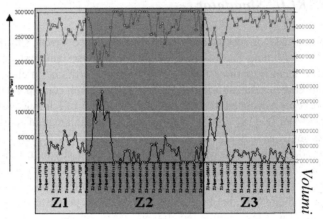

Fig. 8.3 Demand in stores located in three geographical zones (*Z1 hyper* and *super*, *Z2 hyper* and *super*, *Z3 hyper* and *super*)

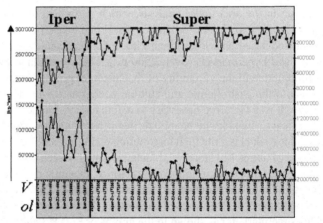

Fig. 8.4 Demand along stores classification: hyper-(*Z1*, *Z2*, *Z3*) and supermarkets (*Z1*, *Z2*, *Z3*)

had a small platform partially serving the local stores. Scenario 1 is based on a regional platform, collecting goods directly from each supplier area and serving all local stores. Scenario 2 is based on the hypothesis of keeping a single large platform and dismissing the other ones; obviously this solution expands into three alternatives characterised by the choice of the platform to be powered as the central one (Zone 1, Zone 2 or Zone 3). Scenario 3 is based on the shuttle hypothesis: each platform receives goods just from the most convenient suppliers and other products through a shuttle service from other platforms; each platform serves just local stores. MARLIN's architecture was based on the idea of interconnecting different modules with specific goals. Blue MARLIN is a stochastic Monte Carlo simulator devoted to evaluating the first ring of the supply chain in terms of flows and costs.

The Virtual Operator is a software agent that generates inquires to web services for populating a database about the travelling paths including details about highway fees, times, distances etc. [20, 21] Sail Fish is a decision support system based on genetic algorithms that identify clusters among the stores in order to optimise the missions in the second ring of the supply chain. Striped MARLIN is a discrete-event stochastic simulator devoted to reproducing in detail all goods transportation and distribution to the stores. Black MARLIN is a model that includes the two previous models (Striped and Blue MARLIN), and a stochastic simulator devoted to calculating the costs related to logistics platform operations for each different management solution (shuttle services, quality controls, etc.). Figure 8.5 presents the overall architecture of the MARLIN system including the different simulators and agents. Blue MARLIN is focused on analysing the infrastructure needs (warehouses, crossdocking platforms) and the central supply chain policies (supplier flow direction, supplier costs and discount policies related to commercial and logistics issues); this is a static stochastic simulation model that estimates the resources required in each node and the best allocation policy for balancing the flows in the first supply chain ring. Striped MARLIN is a discrete-event stochastic simulator that reproduces each single delivery mission as well as the evolution of the store demand based on historical statistical data; by this approach the simulator allows us to evaluate the performance of changing the delivery plan; for instance it is possible to define different sets of multi-drop missions for covering the second ring of the supply chain.

So Striped MARLIN is devoted to evaluating statistically the second ring of the supply chain and allows estimation not only of all direct costs and time performance, but of also the risk of each solution, considering several factors such as:

- *Temperature evolution during the transportation.* Temperature, in fact, is subjected to restrictive regulations and frequent loading/unloading operations due to multi-dropping at different stores introduce a risk of defeating the effectiveness of the freezing system, so the simulator reproduces the temperature behaviour during each mission to quantify the impact of this problem.

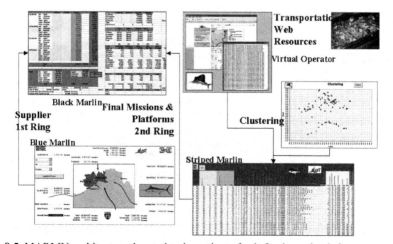

Fig. 8.5 MARLIN architecture devoted to investigate fresh-food supply chain

- *Time gate:* delivery arrival without respect for each store time gate introduces additional costs (if too early it is required to wait, if too late it is required to ask for extra shifts for unloading the truck).
- *Travelling time:* the transportation regulations introduce limits in terms of driving time, and thus mission time, so the simulator is required to measure the risk of exceeding these limits during the mission, introducing additional risks.
- *Capacity variability:* the characteristics of each store in each single mission include a constraint in terms of maximum size of acceptable vehicle; where the demand exceeds the maximum capacity of the largest allocable truck it is required to reorganise the mission (splitting the mission or allocating part of the delivery to other missions) and evaluate the extra costs.

The last element is to summarise the performance of the different logistics solutions and to identify the most reliable and effective alternative in terms of operatives, transportation costs, risk of penalties and extra costs, as well as considering potential discounts from suppliers due to the supply chain reorganisation [22]; this action is in the charge of Black MARLIN, which guarantees coordinated results between Striped and Blue MARLIN. The synthesis of the reports from this component is summarised in Fig. 8.6.

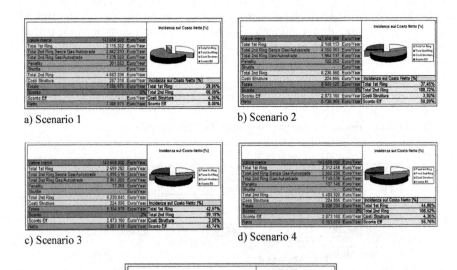

Fig. 8.6a–e Scenario comparison reports

The simulation model was validated by applying mean square pure error evolution analysis, as presented in the graphs of Fig. 8.7a and b. Analysis of Variance (ANOVA) supported successfully validation and verification of MARLIN simulators.

Sail Fish resulted in an optimisation system based on genetic algorithms that allows us to identify the optimal clustering for the stores; this package optimises the missions of the trucks on the second ring of the supply chain. The graph of Fig. 8.8 presents the clustering generated by Sail Fish over the region.

The optimiser operates on a multi-variable target function based on weights related to different costs and risks (truck overloading, extra time, premature arrival, temperature alert levels, etc.); so this fitness function depends on the store clustering and is minimised by the optimiser.

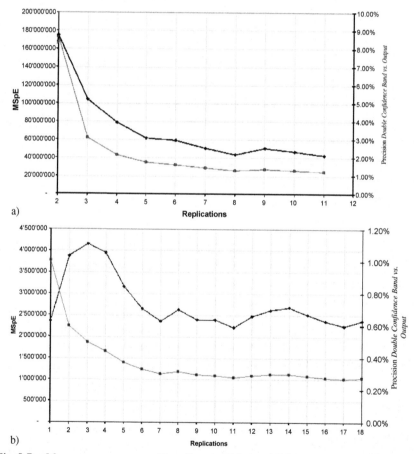

Fig. 8.7 **a** Mean square pure error. Overall cost validation. **b** Mean square pure error. Shuttle cost validation

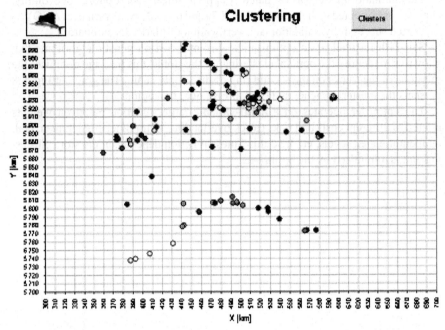

Fig. 8.8 Clustering stores for missions

8.4 Conclusions

These approaches and examples confirm the effectiveness of M&S in this context and the importance of simulating problems where so many parameters affect
the overall performance of the supply chain. Proposed cases related to the supply chains of fresh fish and meat provided an interesting opportunity to test these
methodologies and to develop decision support systems [23] based on optimisation
techniques; the results obtained prove to be very interesting for the sector, while
the analysis outlines the costs/benefits of the different logistics alternatives in each
case. It is possible to acquire additional details on the proposed subjects at:

* http://st.itim.unige.it/cs/logistics
* http://logi1.itim.unige.it/moodle/login/index.php

8.5 Questions

1. Define the advantages/disadvantages of logistics platforms for fresh-food
   distribution.
2. Define a KPI (key performance indicator) for logistics nodes in food distribution.

3. Identify possible logistics solutions for extending fresh-food distribution over a wider geographical area.
4. Evaluate alternative logistics solutions in terms of quality and safety controls.
5. Provide a rational description for trade-off analysis related to introducing a shuttle service among distribution centres.
6. Provide a list of variables to be included in a simulation model for direct/indirect distribution of fresh food.
7. Describe the risk factors related to the distribution missions for fresh food in a network of supermarkets.
8. Describe the constraints that characterise stores, from a logistics point of view, in the goods supply chain.
9. Define a fitness function for optimising the store clustering for food distribution in a retail network.

References

[1] Massei M (2006) Logistics and process solution for supply chain of fresh food in retail. In: Proceedings MAS2006
[2] Bruzzone AG, Genco P, Bocca E et al (2005) Development of a model for evaluating comparative evolution of logistics in retail. In: Proceedings SCSC'05
[3] Bruzzone AG, Massei M, C B (2006) Optimising retail logistics from store point of view. In: Proceedings HMS2006
[4] Bruzzone AG, Viazzo S, Longo F et al (2004) Simulation and virtual reality to modelling retail and store facilities. In: Proceedings SCSC2004
[5] Bruzzone AG (2002) Supply chain management. In: Proceedings Simul 78(5):283–337
[6] Bruzzone AG, Massei M, Brandolini M (2006) Simulation based analysis on different logistics solutions for fresh food supply chain. In: Proceedings SCSC2006
[7] Bruzzone AG, Kerckhoffs E (1996) Simulation in industry. In: 8th European simulation symposium, vols I & II
[8] Bruzzone AG, Bocca E, Massei M et al (2004) Innovative organization solutions and new enabling technologies for fresh meat supply chain management. In: Proceedings MAS2004
[9] Bruzzone AG, Simeoni S, Bocca E (2004) Intelligent management of a logistics platform for fresh goods. In: Proceedings SCI2004
[10] Birkin M, Clarke G, Clarke M (2002) Retail intelligence and network planning. John Wiley, New York
[11] Bruzzone AG, Mosca R, Spirito F et al (2001) M&S for customer satisfaction in retail warehouse management. In: Proceedings EUROSIM2001
[12] Di Conza A, De Michelis S, Barcucci V et al (2003) Artificial neural networks as support for scheduling in food industry. In: Proceedings MAS2003
[13] Bocca E, Briano C (2005) Optimising inventory management in retail supply chain. In: Proceedings SCSC2005
[14] Bruzzone AG, E Page, A Uhrmacher (1999) Web-based modelling & simulation. In: Proceedings SCS International
[15] Ortega B (2000) In Sam we trust. Times Books, New York
[16] Bruzzone AG, Viazzo S, Massei M (2005) Computational model for retail logistics. In: Proceedings WMSCI
[17] Bruzzone AG, Williams E (2004) Modeling and simulation methodologies for logistics and manufacturing optimisation. Simulation 80(3):119–174
[18] Poggi S, Pierfederici L (2007) MARLIN. Technical report ICAMES2007

[19] Bruzzone AG, Mosca R, B C et al (2000) Models for supporting customer satisfaction in retail chains. In: Proceedings HMS2000

[20] Bruzzone AG, Mosca R (1999) Modelling & simulation and ERP systems for supporting logistics in retail. In: Proceedings ESS99

[21] Bruzzone AG, Mosca R, Brandolini M et al (1999) DICO-SAP: Supervising the SAP R/3 retail implementation. COOP Liguria technical report

[22] Bruzzone AG, Viazzo S, Longo F et al (2004) Simulation and virtual reality to modelling retail and store facilities. In: Proceedings SCSC2004

[23] Bruzzone AG, Simeoni S, Bocca E (2004) Intelligent management of a logistics platform for fresh goods. In: Proceedings SCI2004

Chapter 9
Warehouse Order Picking Process

Y. Merkuryev, A. Burinskiene and G. Merkuryeva

Abstract The order picking process – the retrieval of products from specified locations according to customer orders – is the most laborious and costly process in a warehouse. It consumes almost 60% of all warehouse labour activities. Various routing methods can lead to significant improvements. This chapter analyses the influence of routing methods on picker travel distance in a wide-aisle warehouse. In order to determine potential travel distance savings, a simulation model was created. Routing methods in a wide-aisle warehouse and other order picking process optimisations are analysed through simulation. The presented results show that by using appropriate combination of optimisation methods, the picker travel distance can be reduced by about 60%.

9.1 Introduction

Nowadays, a critical topic in warehouse management practice is finding ways to answer the question 'How to increase order picking productivity and how to be more efficient?'. This educational research pays special attention to a customer order picking process with the aim of improving warehouse operational efficiency. Order picking is the retrieval of products from specified locations on the basis of customer orders. The order picking process is the most laborious of all warehouse processes. It may consume almost 60% of all labour activities in the warehouse [1].

Yuri Merkuryev and Galina Merkuryeva
Riga Technical University, Latvia
merkur@itl.rtu.lv, gm@itl.rtu.lv

Aurelija Burinskiene
Vilnius Gediminas Technical University, Lithuania
Aurelija.Burinskiene@vv.vgtu.lt

In this case study, a manual picking process is analysed. An order picker receives a pick list at a computer station, moves to certain picking locations to retrieve products according to the pick list, delivers them to a drop-off point and then moves to a computer station to confirm the completed order and quantities of delivered products [2].

9.2 Objectives of the Project

To study the order picking process, warehouses with narrow aisles that allow the picker to retrieve products from both sides of the aisle are often chosen. However, narrow aisles do not support more than one picker working in the same area. In contrast, this case study, aiming to analyse picker travel distance, examines a warehouse with wide aisles.

The target of the case study is to define combinations of various strategies, by which the minimisation of travel distance can be achieved, thus accelerating the picking process itself.

9.3 Description of Order Picking Process

Order picking is the retrieval of products from specified locations according to customer orders. An order picker always starts the route at a depot. In changing aisles, the picker moves in the direction of the closest cross-aisle. The routing algorithm chooses the shortest way for each aisle individually: the picker needs to return back to the front cross-aisle or to cross the aisle through its entire length to the rear cross-aisle [3, 4]. In a warehouse with two storage blocks, the middle cross-aisle operates as the rear cross-aisle for the first storage block and as the front cross-aisle for the second storage block. This route is called a composite route and is displayed in Fig. 9.1. The composite route is not used very often by other authors.

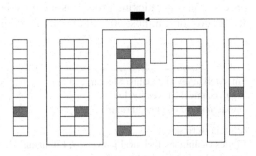

Fig. 9.1 Composite route *(grey* specified locations, *black* depot*)*

Seeking to improve the picking process efficiency, the case study analyses different strategies including:

- Warehouse layout
- Storage strategies
- Customer orders
- Routing methods in a wide-aisle warehouse

The aim of this case is to minimise the picker travel distance in a reference warehouse. While two types of travel distances for order picking are used in literature: average travel distance per order and total travel distance (for a set of orders), the case study focuses on the latter. By minimising the average travel distance, the total travel distance is also minimised [5].

9.3.1 Warehouse Layout

A schematic picture of the reference warehouse with multiple aisles is shown in Fig. 9.2. The width of aisles is often from 2.3 metres to 3.5 metres. Items are stored on shelves and are usually picked in cases. The considered warehouse, called a reference warehouse, has eight parallel aisles, and each aisle contains one hundred picking locations. The cross-aisle in the middle of the warehouse separates it into

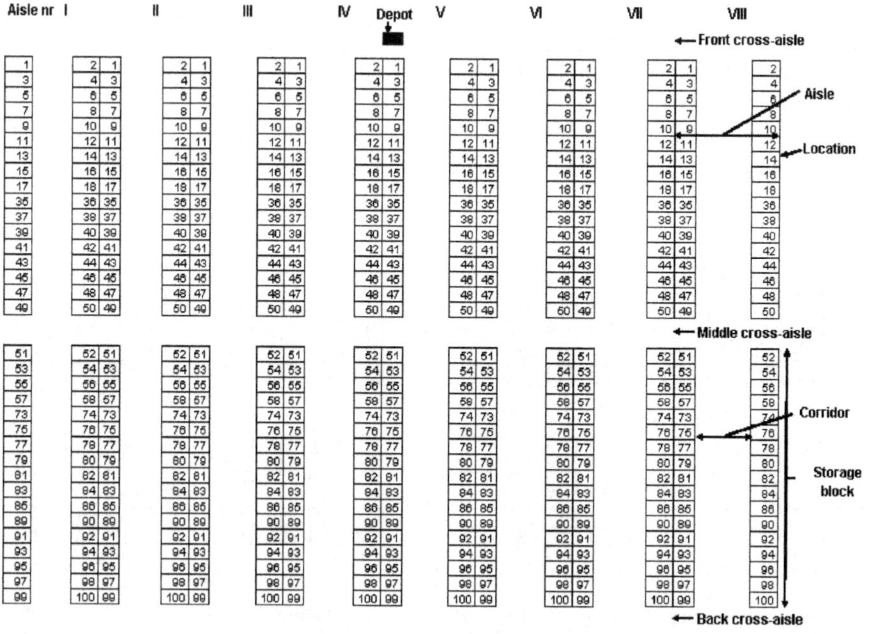

Aisle nr I   II   III   IV   Depot   V   VI   VII   VIII

←— Front cross-aisle

Top block (above Middle cross-aisle) — labels: Aisle, Location

I	II-L	II-R	III-L	III-R	IV-L	IV-R	V-L	V-R	VI-L	VI-R	VII-L	VII-R	VIII
1	2	1	2	1	2	1	2	1	2	1	2	1	2
3	4	3	4	3	4	3	4	3	4	3	4	3	4
5	6	5	6	5	6	5	6	5	6	5	6	5	6
7	8	7	8	7	8	7	8	7	8	7	8	7	8
9	10	9	10	9	10	9	10	9	10	9	10	9	10
11	12	11	12	11	12	11	12	11	12	11	12	11	12
13	14	13	14	13	14	13	14	13	14	13	14	13	14
15	16	15	16	15	16	15	16	15	16	15	16	15	16
17	18	17	18	17	18	17	18	17	18	17	18	17	18
35	36	35	36	35	36	35	36	35	36	35	36	35	36
37	38	37	38	37	38	37	38	37	38	37	38	37	38
39	40	39	40	39	40	39	40	39	40	39	40	39	40
41	42	41	42	41	42	41	42	41	42	41	42	41	42
43	44	43	44	43	44	43	44	43	44	43	44	43	44
45	46	45	46	45	46	45	46	45	46	45	46	45	46
47	48	47	48	47	48	47	48	47	48	47	48	47	48
49	50	49	50	49	50	49	50	49	50	49	50	49	50

←— Middle cross-aisle

Bottom block (below Middle cross-aisle) — labels: Corridor, Storage block

I	II-L	II-R	III-L	III-R	IV-L	IV-R	V-L	V-R	VI-L	VI-R	VII-L	VII-R	VIII
51	52	51	52	51	52	51	52	51	52	51	52	51	52
53	54	53	54	53	54	53	54	53	54	53	54	53	54
55	56	55	56	55	56	55	56	55	56	55	56	55	56
57	58	57	58	57	58	57	58	57	58	57	58	57	58
73	74	73	74	73	74	73	74	73	74	73	74	73	74
75	76	75	76	75	76	75	76	75	76	75	76	75	76
77	78	77	78	77	78	77	78	77	78	77	78	77	78
79	80	79	80	79	80	79	80	79	80	79	80	79	80
81	82	81	82	81	82	81	82	81	82	81	82	81	82
83	84	83	84	83	84	83	84	83	84	83	84	83	84
85	86	85	86	85	86	85	86	85	86	85	86	85	86
89	90	89	90	89	90	89	90	89	90	89	90	89	90
91	92	91	92	91	92	91	92	91	92	91	92	91	92
93	94	93	94	93	94	93	94	93	94	93	94	93	94
95	96	95	96	95	96	95	96	95	96	95	96	95	96
97	98	97	98	97	98	97	98	97	98	97	98	97	98
99	100	99	100	99	100	99	100	99	100	99	100	99	100

←— Back cross-aisle

Fig. 9.2 Wide-aisle warehouse layout

two storage blocks and allows three possibilities to switch between aisles: at the front, at the rear and in the middle.

Products are picked from ground locations. According to the schematic picture of the warehouse, one cell represents one location. Locations are 1.2 metres wide and 0.8 metres deep.

In the case study, a computer station and a drop-off point are represented by the depot. The location of the depot, where the picker starts and ends picking, can be freely chosen by the user before performing simulation.

The explored simulation model is flexible. The depot location is in the middle of the front side of the reference warehouse. The model allows one to change the location of the depot and to study other possibilities (which can be studied without schematic changes):

- The depot can be located (according to Fig. 9.2) at the left or right side, or in the middle of the rear, front or right side of the warehouse (at the left side of the warehouse, a wall is placed).
- The middle cross-aisle can be removed and the number of storage blocks can be reduced.
- The number of aisles can be reduced. It can be decreased from any side of the warehouse, if the number of picking locations for the retrieval in those aisles is equal to zero.

9.3.2 Storage Strategies

In the literature, storage strategies are classified into [5, 6]:
- Random storage – all empty locations have equal probability of being filled.
- Storage in the closest open locations to the depot.
- Dedicated storage – location is reserved for a product, even if the product is out of stock.

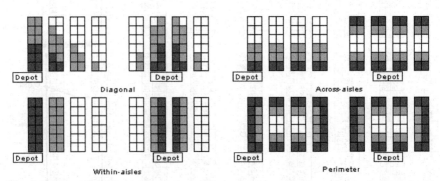

Fig. 9.3 Volume-based storage methods (*dark grey* A products, *medium grey* B products, *white* C products)

- Volume-based storage (can also be called turnover-based storage) – products with the highest sales rates located at the closest to the depot locations (see Fig. 9.3).
- Class-based storage (can also be called 'family-products-based storage') – classes are assigned to a dedicated area of the warehouse. Essentially, it is a combination of random and volume-based storages. In the third-party warehouses, products can be placed by customers and by classes.

Definitions of volume-based storage methods (Fig. 9.3) [4, 7]:
- *Diagonal storage method.* The highest-volume products are located closest to the depot and the lowest-volume products are located farthest from the depot.
- *Within-aisles storage method.* The highest-volume products are located in the aisles closest to the depot and the lowest-volume products are stored in the aisles farthest from the depot.
- *Cross-aisles storage method.* The highest-volume products are located along the front aisles and the lowest-volume products are located along the rear aisles.
- *Perimeter method.* The highest-volume products are located around the warehouse perimeter; the lowest-volume products are placed within the middle of the aisles.

In the case study, products are not assigned to the locations and storage strategies can be used in the following way:
- For example, for volume-based storage methods, a warehouse can be separated into location zones A, B or C, which means that the number of picker visits to locations in zone A is the highest, and locations in zone B are visited by the picker more often than locations in zone C.
- Alternatively, the class-based storage can be mixed with the volume-based one. For example, an aisle can be dedicated to a class, while inside it two location zones can be created in order to assign products according to volume groups: fast movers and middle movers of the class can be placed on one side of the aisle (which represents the first zone), and slow movers and no movers [8] of the class on the other side of the aisle (which represents the second zone). For a product distribution according to volume groups, the authors propose to use the 80/20 rule ('20% of products in the class constitute 80% of picking volume', Fig. 9.4).

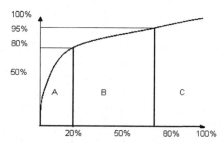

Fig. 9.4 The 80/20 rule

Using our model, the user is free to choose any storage strategy, but he needs to indicate which locations the picker is going to visit.

9.3.3 Customer Orders

A customer order is defined as an order placed by the customer name and date. The customer order can be transferred into one pick list or split into separate pick lists. Such a picking method is called 'picking by order' [5]. However, the smaller customer orders or pick lists are, the more often the picker is returning to the depot and his total travel distance additionally increases. Reduction of the travel distance is more significant when the picker is visiting more locations (the pick list is larger); only the order picking process is taking longer. Customer orders are split into pick lists, based on the customer name, date and the maximum number of picks in the pick list. The maximum number of picks is related to capacity of a picking trolley. When the customer order is large and exceeds the maximum number of picks, it is split into separate pick lists. In the case study, the user has to choose the maximum number of picks.

9.3.4 Routing Methods in a Wide-Aisle Warehouse

In Fig. 9.5, five routing methods in a wide-aisle warehouse are shown.

Each method shows a route in which the picker is expected to move from the current location to the neighbouring one according to the pick list.

The description of each method (according to Fig. 9.5) is given below:
1. The first method (Fig. 9.5, I): first, all locations on one side of the aisle are visited; later the picker visits all locations on the other side.
2. The second method (Fig. 9.5, II): after visiting two locations on one side of the aisle, the picker moves to the other side of the aisle one location up or down (depends on the direction). Each location, according to the logic of this method,

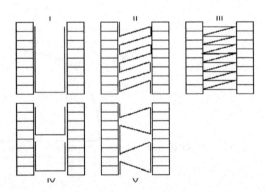

Fig. 9.5 Routing methods in a wide-aisle warehouse

receives an identification number. Identification numbers are used for directing the picker inside the aisle.

3. The third method (Fig. 9.5, III): when a location has been visited on one side of the aisle, another parallel location is visited on the other side of the aisle.

4. The fourth method (Fig. 9.5, IV): after visiting four locations on one side, the picker moves to the other side and visits four parallel locations. Then the picker returns.

5. The fifth method (Fig. 9.5, V): after four locations are visited on one side of the aisle, the picker moves to the other side of the aisle one location back.

In the case study, the above five routing methods in a wide-aisle warehouse are integrated with the composite route and can also be used as a part for other warehouse layout, storage strategies and pick list size tests.

Caron, Marchet and Perego [9] study routing methods in a wide-aisle warehouse. In their study, the routing distance is based on a centreline for travelling along the aisle and on a zigzag travelling inside the aisle to retrieve products from both sides of the aisle. The zigzag travelling method is illustrated in Fig. 9.6. Here one pick list is processed. All locations specified for this picking are not placed in front of each other, and therefore it is difficult to use methods from Fig. 9.5.

In order to define the most efficient routing method in a wide-aisle warehouse and to analyse possibilities to optimise the order picking process, a simulation model for the reference warehouse was created.

9.4 Model Description and Instructions

For the case study, an Excel simulation model was created that consists of the following parts:

- Warehouse layout (sheet 'DC') and location names database (sheet 'Location names')
- Location visit identification numbers database (sheets 'Location visit ID', '4_5th method')
- Pick lists database (sheet 'Pick lists')
- Simulation on warehouse layout algorithm (sheet 'Picking')

Efficiency of the picking process is measured in the model by the picker travel distance, the number of picked orders and the number of locations visited.

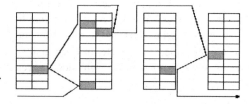

Fig. 9.6 Zigzag travelling in the wide-aisle

9.4.1 Warehouse Layout and Location Names Database

A. The warehouse layout, sheet 'DC', is flexible regarding changes according to the following instructions:
   - *Changes in the depot location.* The depot can be located on the left or right side or in the middle of the rear or the front, or on one side of the warehouse.
   - *Changes in the middle cross-aisle.* Simulation of the picking process without the middle cross-aisle can be performed.
   - *Changes in the number of aisles.* The number of aisles can be decreased for the simulation from the right side of the warehouse.
B. Location names database, sheet 'Location names'. Each location name consists of the aisle name and of the location place number inside the aisle. The aisle name is based on a Roman numeral value and the place number is based on an Arabic numeral value. For example, the full location name is 'VII-50'. The full names of all locations are displayed in the sheet 'Location names'. This is a technical sheet, which cannot be deleted or modified.

9.4.2 Location Visit Identification Numbers Database

Each of 800 locations (eight aisles with one hundred locations each) of the reference warehouse received an identification number in the sheets 'Location visit ID' and '4_5th method'.

The identification number is unique to a particular location and is based on a routing method in a wide-aisle warehouse (see Fig. 9.5). The location visit identification numbers database was created based on five picker routing methods in a wide-aisle warehouse (as described in Sect. 9.3.4) which were implemented with some modifications for the reference warehouse. As a result, sets of identification numbers were created for routing purposes, in accordance with different routing methods.

The first routing method. When all locations on the right side in any aisle are visited, the picker moves from the front to the rear of the warehouse, then he moves to the left side (according to the picker) and locations on the left side of any aisle are visited; finally the picker returns. In some specific situations, this routing method can change the composite route into the return route. (The logic of the return route is as follows: the aisle is exited from the same side as entered.) Identification numbers are created according to the described logic.

The second routing method. The picker starts picking on the right side of any aisle and finishes picking on the left side of any aisle. In odd aisles he starts at the lowest location number in the right side and finishes at the highest location number in the left side of the aisle, in even aisles the reverse. Inside the aisle, the picker visits two locations on the right side, moves to the left side one location back, visits two locations on the left side, moves to the right side one location ahead. Identification numbers are created according to the described logic:

1. Identification numbers for each fourth location (1, 5, 9, etc.) in the first aisle are equal to a location number.
2. Identification numbers for the first four locations in other aisles are calculated individually:
   - For four front locations in odd aisles by:

$$1\text{st location}=(\text{number of aisle}-1)\times 100+1, \qquad (9.1)$$

$$3\text{rd location}=(\text{number of aisle}-1)\times 100+2, \qquad (9.2)$$

$$2\text{nd location}=(\text{number of aisle}-1)\times 100+3, \qquad (9.3)$$

$$4\text{th location}=(\text{number of aisle}-1)\times 100+4. \qquad (9.4)$$

   - For four rear locations in even aisles using:

$$99\text{th location}=(\text{number of aisle}-1)\times 100+1, \qquad (9.5)$$

$$97\text{th location}=(\text{number of aisle}-1)\times 100+2, \qquad (9.6)$$

$$100\text{th location}=(\text{number of aisle}-1)\times 100+3, \qquad (9.7)$$

$$98\text{th location}=(\text{number of aisle}-1)\times 100+4. \qquad (9.8)$$

   - For all other locations in odd aisles:

$$\text{ID number of location } n=\text{ID number of location }(n-4)+4 \qquad (9.9)$$

   (in 'n − 4', '4' represents four sequential locations).
   - For all other locations in even aisles, formula (9.9) can be used, but locations for determining identification numbers have to be sorted in descending order.

The third routing method. The picker starts and finishes picking in odd and even aisles as described in the second method. The picker picks product from one location on the right side of the aisle and moves to pick product from a parallel location on the left side of the aisle. Identification numbers are calculated accordingly:

1. For locations in the first aisle they are equal to the location place number.
2. For locations in other odd aisles they are calculated by the following formula:

$$(\text{number of aisle}-1)\times 100+\text{location number}. \qquad (9.10)$$

3. For even aisles:
   - For 99th location, visit identification numbers are calculated using:

$$(\text{number of aisle} - 1) \times 100 + 1. \qquad (9.11)$$

   - For 2nd location, visit identification numbers are calculated as follows:

$$(\text{number of aisle} - 1) \times 100 + 100. \qquad (9.12)$$

4. All other locations in the odd aisle are sorted in ascending order and identification numbers are placed among front and rear locations. The same holds for even aisles, but to determine identification numbers, locations are sorted in descending order (see Fig. 9.7).

The fourth routing method. After four locations are visited on the right side, the picker moves to the left side of the aisle to visit four parallel locations in reverse, then being on the left side of the aisle he moves to pick another four locations and moves to the right side to visit four parallel locations in reverse, and so on. An example of location visit identification numbers, based on the previously described logic, is provided in Table 9.1.

Locations on the right and left sides in the odd aisle are sorted in ascending order separately for each side, and locations on the right and left sides in the even aisle are sorted in descending order separately for each side. After placing identification numbers for the first 16 locations at the beginning of each aisle, identification numbers for other locations are calculated using the following formula:

$$\text{ID number of location } n = \text{ID number of location } (n - 16) + 16 \qquad (9.13)$$

(in '$n - 16$', '16' are 8 locations on one side of the aisle and 8 locations on the other side).
As the number of locations (100) per aisle cannot be divided by 16, the last four locations at the end of the aisle are visited in the following way: two locations are visited on the right side; the picker moves to the left side to visit two parallel locations in reverse, and after that moves to the rear cross-aisle in an odd aisle and to the front cross-aisle in an even aisle.

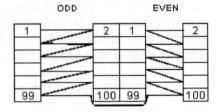

Fig. 9.7 The third routing method: between odd and even aisles and routing inside aisles

The fifth routing method: the picker visits two locations at the beginning of the aisle on the right side and moves to the left side, back one location in the aisle; on the left side he visits four locations and moves again to the right side, back one location. The picker starts each aisle by visiting two locations on the right side and finishes the aisle by visiting two locations on the left side: odd aisles are started at the lowest location place number, even aisles at the highest location place number on the right side of the aisle. Inside the aisle, locations are split into blocks:

1. On the right side, after the first two locations, all other locations are split into blocks of four.
2. On the left side, starting from the first location, all locations are split into blocks of four.

Inside the block, the identification number increases by one, but between blocks on the odd side or between blocks on the even side, the identification number increases by five, as, for example, in Table 9.2.

In the sheet '4_5th method' identification numbers for locations are created according to the fourth and the fifth routing methods in a wide-aisle warehouse.

9.4.3   Pick Lists Database

All customer orders are entered into the sheet 'Pick lists'. It contains the following details: customer name, date and product place (location name). Each customer order created by the user can be directed straight to the order picker by adding location visit identification numbers. Alternatively, large orders (or orders with products

Table 9.1 Location visit identification numbers for the fourth routing method

Location number	Identification number	Location number	Identification number
I-1	1	I-2	8
I-3	2	I-4	7
I-5	3	I-6	6
I-7	4	I-8	5
I-9	16	I-10	9
I-11	15	I-12	10
I-13	14	I-14	11
I-15	13	I-16	12
I-17	ID no. (I-1)+16	I-18	ID no. (I-2)+16
...	...	...	...
I-97	97	I-98	100
I-99	98	I-100	99

Table 9.2 Location visit identification numbers for the fifth routing method

Block number for four locations	Location number	Identification number	Block number for four locations	Location number	Identification number
0	I-1	1	1b	I-2	3
0	I-3	2	1b	I-4	4
1a	I-5	ID no. (I-3)+5	1b	I-6	5
1a	I-7	8	1b	I-8	6
1a	I-9	9	2b	I-10	ID no. (I-8)+5
1a	I-11	10	2b	I-12	12
2a	I-13	15	2b	I-14	13

that are located in different zones (volume-based or class-based storage zones) of the warehouse) can be split up and retrieved by a number of several pick lists (in such cases volume-based or class-based storage zones need to be specified). In the model, the picker is picking products according to pick lists.

The first location in a pick list can be any location from the warehouse. If the storage strategy is volume-based, there is a high probability that the pick list starts at a location from zone A (see Sect. 9.3.2 to review storage strategies). All ranges of locations can be entered into the pick lists by the user or the user can choose locations randomly (Excel function 'Rand' can help to generate a random number for 100 locations and functions 'Roundup' or 'Rounddown' used together can help to round the random value).

How can the pick lists database be created? There are several steps in the following sequence:

- User chooses customers and locations, fixes the date and enters these data into the 'Pick lists' sheet. It is important to mention that the sequence or positions of columns in this sheet cannot be changed.
- User chooses the routing method in a wide-aisle warehouse.
- User chooses the maximum number of picks in the pick list.
- Reference number of each pick list is allocated by a macro command as follows:
  - If a customer order exceeds the maximum number of picks, then it is split into more pick lists, and a reference number for each pick list is allocated.
  - If a customer order does not exceed the maximum number of picks, then it stays as one pick list and a reference number for this pick list according to the pick list sequence is allocated.
- Location visit identification numbers are entered by a macro command.
- Picks in the pick list are sorted (in ascending order) by a macro command based on location visit identification numbers.

9.4.4 Simulation Algorithm

The simulation algorithm is described in the sheet 'Picking'. It uses the pick lists database, where picking tasks and a sequence in which locations shall be visited are defined. For picking in accordance with the next pick list, the picker first returns to the depot [1].

Simulation of the picking process involves movements of the order picker inside aisles, cross-aisles, as well as returns to the depot [1], which are modelled by macro commands in Visual Basic.

During simulation, the current pick list gets its reference number for identification purposes automatically. Such numbers can also be indicated manually by the user, if needed, before simulation, to check which locations are chosen for the specific pick list. If the pick list number is not indicated, the number of visits to each location from the 'pick list' sheet is summarised in a schematic picture of the warehouse (it can be checked how often pick lists for each location are created and how the storage strategy is used). The number of visits can be ranged by different criteria values. For displaying ranged data, the MS Excel Conditional Formatting function is used. Locations coloured dark grey are the most visited; locations coloured light grey are visited more often than locations coloured white. For the proper ranging function, the user has to check which location has the highest number of visits and how big the number of visits is and, after that, to change range values. The best option is to divide the highest number of visits per location into three equal parts, but other possibilities can also be used.

Picker movements are shown in a schematic picture in the sheet 'Picking'. Picking can be started with the macro command button 'Picking'. Generally, during simulation the picker is moving on the right side of the corridor and is stepping to the left side only by a pick list request. The picker is moving according to the schematic picture of the reference warehouse. For simulation, the following changes in the reference warehouse (described above) can be used without changes in the schematic picture: the depot location, middle cross-aisle and number of aisles. The results of each routing method are displayed in separate rows.

Total travel distance calculation. The aisle width is 5 metres; the total width of a corridor is 3.4 metres. These numbers are used for total travel distance calculation. When the picker moves horizontally, the simulation algorithm checks each column type.

Inside the corridor, it is considered that the picker walks 0.3 metres away from the shelves. When the picker is retrieving a product, 0.6 metres (0.3 metres from location and 0.3 metres back) are added to the total travel distance.

If the picker steps from the right side of the corridor to the left side or from the left to the right side, the distance of 2.8 metres for changing sides is added. The length of each location is 1.2 metres. The middle cross-aisle is 4.8 metres wide; it is used in the reference warehouse to handle the traffic.

Simulation algorithm logic. A macro command is used to calculate the total travel distance according to the selected routing method. On average, it takes five minutes to run a simulation for 800 locations and eight pick lists. Using such possibilities, the user can test either one routing method with different layout and pick list size combinations or different routing methods for the same situation.

Three macro command blocks are used for the algorithm:

- The first two command blocks – for moving between even or odd locations.
- The third command block – for returning to the depot.

For simulating picker movements, a blue colour is used. When the picker moves into a new cell in the schematic picture, the colour in the previous cell is removed and the new cell is coloured. When the picker moves across, the differences in rows and columns of previous and current locations (cells) for total distance calculations are taken into account. If the picker moves between different aisles, the algorithm checks which cross-aisle is closer to the previous and next locations, and indicates the closest cross-aisle for the next picker movements.

If the user is willing to run more tests for the efficiency analysis, he has to copy results of the previous simulation.

9.5 Verification and Validation

Verification and validation of the model were performed.

Verification tests:

- Calculation of the total travel distance for picking in a warehouse with one storage block was corrected in the macro command algorithm.
- The depot location was tested with the macro command. In the warehouse with two storage blocks, the depot may even be located in the middle cross-aisle, if the user wishes.
- In reality, when the last pick list is finished, the picker returns to the depot; this point was accordingly corrected in the macro command.
- Unused data was removed from the tables.
- Additional functionality of checking statistical data (the number of visits for each location) was entered into the model; this information can be automatically updated from the 'Pick lists' sheet to the 'Picking' sheet.
- When the picker is moves inside the corridor from the left side to the right side to return to the depot, the total travel distance does not increase.

Validation tests:

- The simulated model meets real processes which occur in an ordinary warehouse. Pick lists were created based on the date, customer, pick lines and location visit identification numbers.
- In reality, many pickers are picking big orders, but the calculation of travel distance implemented in the model represents a summary of all pickers' travel distances.
- In practice, the picker can return to the depot through different cross-aisles or corridors. In that case, the total travel distance of the composite route can be the same or longer, as the macro command algorithm chooses the shortest route.

9.6 Tasks for the Reader

One of the difficulties most often found in warehouse operation is that a product assortment is growing continuously. The increasing product assortment has then to be stored, so the process requires an increasing amount of a floor space. Furthermore, the amount of orders tends to increase and simultaneously the size of each customer order decreases [4] as customers are placing orders more frequently. In such situation, a typical task is to find an optimal size of the pick list.

Other tasks:

- Find advantages of various storage strategies and suggest the best one.
- Share some ideas about the depot location and guidelines regarding which location can be used for a particular situation.
- Create and describe your own routing method in a wide-aisle warehouse.
- Test the necessity of a middle cross-aisle.

Find the best results for individual tasks and later combine some tasks.

The reader is invited to experiment with the above-described simulation model. It is located at http://www.itl.rtu.lv/Case_studies_Chapter_9/.

9.7 Experiments

For the performed experiments, eight pick lists for all 800 locations were created: one pick list per aisle. From the simulation tests we can see that the total travel distance is sensitive to the depot location. These tests confirm that the best location of the depot is in the case of the reference warehouse.

In the simulation case, the first routing method in a wide-aisle warehouse is replacing the return route. As can be seen from Fig. 9.8, the total travel distance

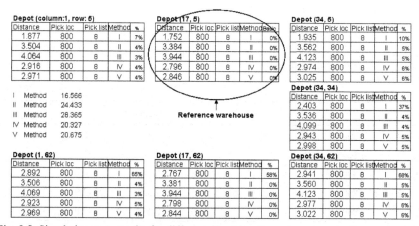

Fig. 9.8 Simulation tests results for different depot locations, in metres

in some tests exceeds 2,000 metres. In Fig. 9.8 combinations of different routing methods and depot locations are compared (as a percentage) with the picker total travel distance in the reference warehouse. In the case of the return route, the total travel distance increases when the picker crosses an aisle through its entire length to start picking and to return to the depot (if the depot is located at the rear in odd aisles or the depot is placed at the front of the warehouse in even aisles).

By summarising the results obtained for each routing method for all tests with the depot location, we can see that the first routing method gives the best results (I: 16,566, II: 24,433, III: 28,365, IV: 20,327, V: 20,675). With the first routing method, the picker travel distance is 2.95 metres per location (for comparison, with the third routing method it is 5.06 metres per location).

From the experiments, we can see that the picker always crosses an aisle through its entire length and does not use the middle cross-aisle. Due to that, the middle cross-aisle is removed in the reference warehouse and the total travel distance is recalculated for the first routing method in a wide-aisle in the reference warehouse. After recalculation, the total travel distance is 1,675 metres, which is 4.4% shorter than the previously obtained 1,752 metres.

The total travel distance is also split into two parts: the zigzag (inside the corridor between right and left sides) and centreline distance (centreline = total travel distance minus zigzag). The centreline distance is also divided into three parts:
- From the depot to the first location in the pick list (distance to start of picking)
- From the last location in the pick list to the depot (distance to end of picking)
- The basic centreline

The resulting total travel distances (in the parts) for the reference warehouse with middle cross-aisle removed are presented in Table 9.3. During different simulation tests, the same locations were visited. In general, starting and finishing locations of a particular pick list depend on location visit identification numbers. However, in *the first pick list* the starting location is always the same.

Sensitivity of the routing method to the depot location was tested. The location of the depot was changed to the middle at one side of the warehouse. Tests with the

Table 9.3 Total travel distances in parts for five routing methods in the reference warehouse without the middle cross-aisle

Distance						Parts of centreline	
In metres	Pick loc	Pick list	Method	Zigzag	Centreline	To start	To depot
1,675	800	8	I	42	1,633	106.8	107.2
3,230	800	8	II	1,117	2,113	342.0	342.4
3,867	800	8	III	2,234	1,633	342.0	341.2
2,642	800	8	IV	311	2,331	342.0	340.0
2,693	800	8	V	580	2,113	342.0	342.4

third routing method were repeated and parts of the total travel distance were evaluated. For each part of the total travel distance, the following results were obtained: for zigzag, 2,234 metres (the same as in the reference warehouse); for returning to the depot, 412.8 metres; for travelling to the first location to start picking, 424.8 metres. From these results it can be concluded that the depot location does not affect the zigzag travel distance (distance inside the corridor between the right and left sides).

For the efficiency analysis, tests with the maximum number of picks for pick lists were performed. For the first routing method in a wide-aisle warehouse (in the reference warehouse without the middle cross-aisle), the following maximum numbers of picks per pick list were chosen: 100, 150 and 200. According to one pick list, 100, 150 and 200 locations were visited and one item was picked from each location. First, the quantity of pick lists was reduced from eight to six, and later, from six to four. The total travel distance results consist of the basic centreline and other parts of the total travel distance. Simulation results are displayed in Fig. 9.9. As we can see, the longer the pick list is, the more effective travel distance becomes: percentages of 'to start' and 'to depot' distances are decreasing in total travel distance.

From these tests we can draw a conclusion that the first routing method in a wide-aisle warehouse does not increase efficiency in all situations. The total travel distance received for 100 picks is 1,675 metres; for 150 picks, 2,012 metres; and for 200 picks, 1,592 metres. Actually the most efficient total travel distance is in the case of 200 locations (two aisles). The results of these tests demonstrate that in case of a new situation the routing method in a wide-aisle warehouse has to be checked again.

At the end of the study, the volume-based storage strategy was implemented. Two zones 'food' and 'non-food' were chosen. The warehouse was split into two equal parts (four aisles per each zone). Inside the zones, the random storage strategy was combined. Six pick lists were constructed (the maximum number of picks was chosen equal to 100, with different pick lists for each zone) and 503 locations were visited. The total travel distance for the first routing method in a wide-aisle warehouse is 1,779 metres, 3.5 metres per location and 296.5 metres per pick list.

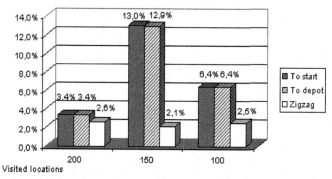

Fig. 9.9 Number of picks: other parts of the total travel distance for the first routing method

9.8   Concluding Remarks

The described experiment demonstrates that in the considered situation the efficiency of the picking process can be improved by maximum 68% by placing the depot properly and choosing the right routing method (see Fig. 9.8).

The target of this case is to have a possibility of evaluating five routing methods in a wide-aisle warehouse at the pick list creation moment in order to create the pick list according to the best one.

For five routing methods in a wide-aisle warehouse, the total travel distance and its parts were calculated through simulation; it was found that:

- The zigzag distance influences the total travel distance from 2% to 57% (the third routing method in a wide-aisle warehouse).
- The zigzag distance increases when the number of picks per aisle is higher, but it is not affected by the depot location.

The concluding remarks for the travel distance are as follows:

- The total travel distance is higher when the number of aisles included in the pick list is odd.
- The difference between all five routing methods is more considerable when the number of locations included in the pick list is higher.

9.9   Questions

1. In the model, a warehouse with two blocks is used. Is it enough to have one middle cross-aisle in the reference warehouse? In which particular situation is the middle cross-aisle needed and helps to reduce the total travel distance?
2. When is the middle cross-aisle necessary:
   - if the maximum number of picks per pick list is relatively low, or
   - if the maximum number of picks per pick list is relatively high?
3. What might be the best depot location when pick lists are received by the picker electronically [2]?
4. How successful might a storage method be when two storage blocks of the reference warehouse are visited separately and successively by the picker?
5. How much does the family-products-based storage strategy influence the picking process in the case of:
   - The warehouse separated into parts?
   - Depot location movement?
   - Picking the same products for different customers?

References

[1] Merkuryeva G, Machado CB, Burinskiene A (2006) Warehouse simulation environments for analysing order picking process. In: Proceedings international mediterrranean modellling multiconference, pp 475-480

[2] Petersen CG, Schmenner RW (1999) An evaluation of routing and volume-based storage policies in an order picking operation. Decis Sci 30(2):481–501

[3] Petersen CG (1997) An evaluation of order picking routeing policies. Int J Oper Prod Manag 17(11):1098–1111

[4] Roodbergen KJ, Petersen CG (1999) How to improve order picking efficiency with routing and storage policies. In: Forger GR et al (eds) Perspectives on material handling practice. Material Handling Institute, Charlotte, NC

[5] De Koster R, Le-Duc T, Roodbergen KJ (2006) Design and control of warehouse order picking: a literature review. Erasmus Research Institute of Management, Erasmus University, Rotterdam

[6] Roodbergen KJ (2001) Layout and routing methods for warehouses. Erasmus Research Institute of Management, Erasmus University, Rotterdam

[7] Dukic G, Oluic C, Cala I (2002) Order-picking: largest gap routing policy with COI-based storage policy. In: Proceedings 18th international conference on CAD/CAM, Robot Fact Future, pp 49–55

[8] Saenz N (2002) Order picking operations design. In: Forger R et al (ed) Perspectives on material handling practice. Material Handling Institute, Charlotte, NC

[9] Caron F, Marchet G, Perego A (2000) Layout design in manual picking systems: a simulation approach. Integr Manuf Syst 11(2):94–100

Chapter 10
Material Handling System

G. Neumann

Abstract A highly complex material handling system for pallets forms the interface and link between a warehouse, two production areas and an order-picking area of a company handling numerous crossing flows of palletised raw materials, products, packaging materials etc. Simulation-based analysis of the system aims at investigating its performance to determine the system's load limit and derive conclusions regarding the system's ability to cope with future loads. The case study presented in this chapter extracts key aspects from the respective real project to show how such a problem typical to logistics is approached and solved.

10.1 Objectives of the Project

The main objective of the project is to analyse the flows and system performance for:
- Estimating the maximum load the material handling system could cope with because of its design and technical parameters
- Analysing bottlenecks hindering system performance at load limit
- Deriving recommendations for improving system performance without changing its design or elements

The material handling system's workload limit is estimated from the number of pallets per hour that are concurrently provided to and removed from each of the production and order-picking areas. The potential of conflicts resulting from this 'competition' for resources finally determines a performance limit below the sys-

Gaby Neumann
Otto-von-Guericke-Universität Magdeburg, Germany
gaby.neumann@ovgu.de

Y. Merkuryev et al. (eds.), *Simulation-Based Case Studies in Logistics*
© Springer 2009

tem's technical capability. Therefore the project aims at proportionally balancing flows in such a way that this loss of performance is minimised.

10.2 Description of the Material Handling System

This section describes the material handling system to be analysed with its components, parameters and flows relevant for model building and simulation.

10.2.1 System Functionality

The material handling system interlinks a company's warehouse (high-bay store), two production areas (manufacturing, packaging) and an order-picking area (via a depalletising robot). It is passed by several types of palletised goods (e.g. boxes, barrels) in crossing flows: the warehouse holds refilling stocks, work-in-progress stocks, raw material stocks, etc. necessary to provide the order-picking area with ready-made products to serve customer orders. The lifecycle of such products usually starts at the warehouse from where raw materials are moved to the manufacturing area. After manufacturing, products are stored in the warehouse until there is a respective demand for them. Exceptionally, certain products might also be forwarded to the depalletising robot directly. Then they are removed from the warehouse and provided to the packaging area. Packaging of products aims at making them ready for delivery to customers. Therefore pallets from packaging are usually forwarded directly to the depalletising robot. Alternatively, ready-made products might also be stored in the warehouse. Finally, empty pallets leaving the depalletising robot need to be returned to the warehouse for further use.

10.2.2 System Structure and Boundaries

The system is linked to its environment by a number of transfer points where pallets enter or leave the system. These transfer points form the sources and sinks of the system to be investigated (see Fig. 10.1). Whereas the links to the warehouse (IN/OUT_WH) work automatically and without any significant time delay, providing and receiving of pallets at transfer points of the two production areas (OUT/IN_PACK and OUT/IN_MAN) are operated manually with average handling times of 2 min per pallet. In contrast, the link to the order-picking area is not classified as system boundary because here only depalletising operations are carried out with the resulting empty pallets not leaving the system.

Pallets are of standard size (800 × 1,200 mm) and are moved by use of automated mechanical handling technology, i.e. non-accumulating chain conveyors (CC),

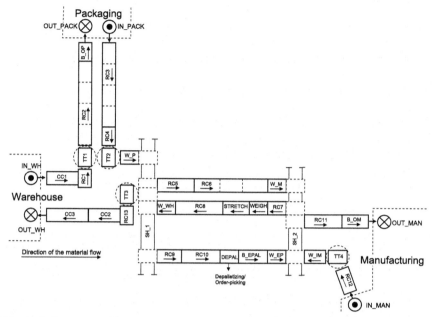

Fig. 10.1 Structure and boundaries of the material handling system

accumulating roller conveyors (RC), turntables (TT) and shuttles (SH). Except for those components for which Fig. 10.1 shows intersections separated by dotted lines (RC2, RC3 and RC6), all system components have a handling capacity of one pallet at a time. Further technical parameters have been identified as follows:

- Standard conveying speed of all technological components is 0.17 m/s.
- A 90° turn of a turntable takes about 6 s independently of its loading state.
- Shuttles move at a travelling speed of 0.4 m/s independently of their loading state, but slow down to 0.1 m/s for the last 10 cm of the travelling distance.

Pallets targeted to the order-picking area usually contain two stacks of eight boxes each for refilling order-picking buffers. A depalletising robot (DEPAL) removes these boxes from the pallets in pairs and puts them on a belt conveyor that moves the boxes to the order-picking area. The total depalletising time for completely emptying a pallet is about 140 s. Empty pallets are forwarded to the attached buffer of empty pallets (B_EPAL) where they are stacked. As soon as a stack reaches a height of nine pallets its movement back to the warehouse is started.

10.2.3 Pallet Flows

Pallet flows in the system usually serve unambiguous source–sink relations according to a pallet's pre-defined and known target destination (see Fig. 10.2).
*   Warehouse – Packaging
    (IN_WH → OUT_PACK: empty pallets, packaging materials, products)
*   Warehouse – Depalletising robot
    (IN_WH → DEPAL: ready-made products for order-picking)
*   Packaging – Depalletising robot (or Warehouse) (IN_PACK → DEPAL
    (or OUT_WH): ready-made products for order-picking)
*   Depalletising robot – Warehouse (DEPAL → OUT_WH: stacks of empty pallets)
*   Warehouse – Manufacturing (IN_WH → OUT_MAN: raw materials, bottles, etc.)
*   Manufacturing – Warehouse (IN_MAN → OUT_WH: products)

Exceptions to these 1:1-relations form the flows from the two production areas, each of which might split up towards two targets: some pallets from packaging are not forwarded straight to the depalletising robot, but stored in the warehouse instead (see Fig. 10.2c). Those pallets have to be checked for their overall dimensions at the weighing and stretching stations (WEIGH, STRETCH) first and therefore need to be delivered by shuttle SH1 to roller conveyor RC5. From there the pallets move via roller conveyors RC6 and B_OP, shuttle SH2 and roller conveyors RC7, WEIGH, STRETCH, RC8 and W_WH back to shuttle SH1, by which they are finally moved to the warehouse accessing conveyors.

Instead of being delivered to the warehouse for interim storage, pallets from manufacturing might eventually be forwarded directly to the depalletising robot to immediately be available in the order-picking area (see Fig. 10.2f). In this case shuttle SH1 provides pallets to the roller conveyor RC9 instead of warehouse accessing conveyors.

10.2.4 Process Control

Specific rules for process control apply to the two shuttles, but also to the transfer of pallets between neighbouring roller conveyors:
*   A shuttle is called by a pallet when the pallet is in place at any of the shuttle's entrance points.
*   If there is more than one call for shuttle SH1, priority is given to pallets from manufacturing to the depalletising robot. Calls from all other pallets are handled according to their arrival (FIFO principle).
*   Shuttle SH2 always handles stacks of empty pallets waiting at W_EP first to avoid any traffic jam behind the depalletising robot, since this could hinder or even stop provision for the order-picking area. Second priority is given to pallets from manufacturing to the warehouse waiting at W_M. Calls from all other pallets are handled according to their arrival (FIFO principle).

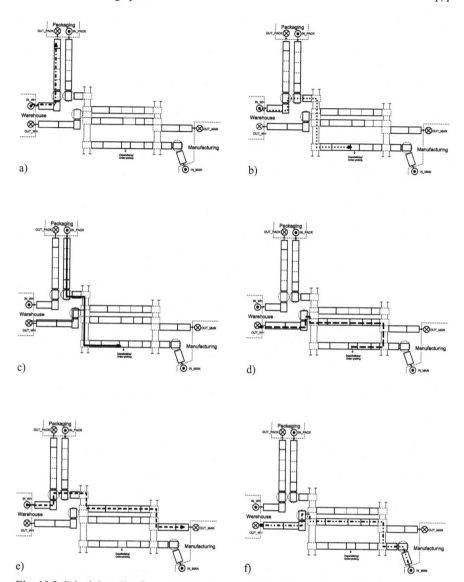

Fig. 10.2 Principle pallet flows through the material handling system. **a** Warehouse – Packaging. **b** Warehouse – Depalletising robot. **c** Packaging – Depalletising robot. **d** Depalletising robot – Warehouse. **e** Warehouse – Manufacturing. **f** Manufacturing – Warehouse

- For pallet movement on or between roller conveyors the conveying strategy of forward control applies. This means a pallet is only moved from one conveyor or conveyor section to its successor if there is enough capacity for the pallet as a whole.

10.3   System Analysis

Since the project aims to determine the system's load limit, i.e. the maximum amount and frequency of pallets which can run through the system in a certain amount of time, pallet flows according to the push principle are assumed. In other words, all transfer points where pallets enter the system (sources) always deliver a pallet when its succeeding conveyor has free capacity. Consequently, the total number of incoming pallets directly depends on the maximum possible performance (in terms of throughput) of the mechanical handling technology they have to pass on their way through the system.

To derive a first estimate of the system's load limit measured in pallets per hour the maximum hourly throughput for each individual relation (see Sect. 10.2.3) is calculated on a flow-by-flow approach without taking into consideration any flow interferences (i.e. each flow exclusively uses all necessary resources). Since all flows follow their individual paths through the system (see Fig. 10.2), the critical system component limiting a flow varies between the relations. For determining the technical load limit per relation these critical components have to be identified and their performance maximum has to be calculated. The results of this flow-specific analysis are summarised in Table 10.1.

The system's theoretical load limit results from superposing the individual flows (see Fig. 10.3) considering possible concurrent use of joint resources. The latter

Table 10.1 Maximum throughput per pallet flow (exclusive use of resources)

Pallet flow	Critical system component	Maximum hourly throughput
Warehouse – Packaging	OUT_PACK (manual operation)	$\dfrac{3{,}600 \text{ s/h}}{120 \text{ s/pal}} = 30 \text{ pal/h}$
Warehouse – Depalletising robot	DEPAL (depalletising operation 140 s + changing of pallets 8.24 s)	$\dfrac{3{,}600 \text{ s/h}}{148.24 \text{ s/pal}} = 24.28 \text{ pal/h}$
Packaging – Depalletising robot	IN_PACK (manual operation)	$\dfrac{3{,}600 \text{ s/h}}{120 \text{ s/pal}} = 30 \text{ pal/h}$
	DEPAL (depalletising operation 140 s + changing of pallets 8.24 s)	$\dfrac{3{,}600 \text{ s/h}}{148.24 \text{ s/pal}} = 24.28 \text{ pal/h}$
Depalletising robot – Warehouse	DEPAL (generation of empty pallets)	$\dfrac{24.28 \text{ pal/h}}{9 \text{ pal/stack}} = 2.7 \text{ stack/h}$
Warehouse – Manufacturing	OUT_MAN (manual operation)	$\dfrac{3{,}600 \text{ s/h}}{120 \text{ s/pal}} = 30 \text{ pal/h}$
Manufacturing – Warehouse	IN_MAN (manual operation)	$\dfrac{3{,}600 \text{ s/h}}{120 \text{ s/pal}} = 30 \text{ pal/h}$

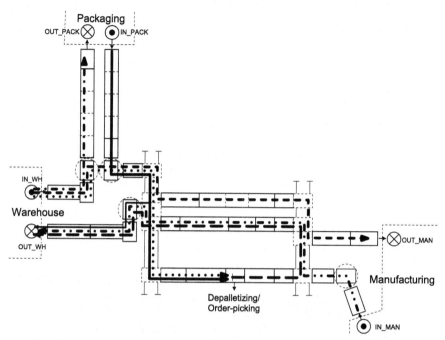

Fig. 10.3 Use of resources by superposed pallet flows

can only be found with the depalletising robot (DEPAL). Since it is the critical component for two different flows (IN_WH → DEPAL and IN_PACK → DEPAL), its performance maximum of 24.28 pal/h is to be understood as the total number of pallets per hour for both flows together. Assuming a maximum load with all relevant components (sources, sinks and depalletising robot) the possible performance maximum of the system as a whole can now be calculated as follows:

IN_WH → OUT_PACK	30.0 pal/h
IN_WH → DEPAL + IN_PACK → DEPAL	24.3 pal/h
DEPAL → OUT_WH	2.7 stacks/h
IN_WH → OUT_MAN	30.0 pal/h
IN_MAN → OUT_WH	30.0 pal/h
Possible maximum system load	117.0 objects/h (pallets/stacks)

This theoretical value resulting from analytical thoughts is to be understood as the absolute performance limit. Due to randomly varying manual operation times and situation-specific concurrency in the use of resources, leading to delays in pallet flows and additional waiting times of pallets that can hardly be calculated, the real load limit can be expected to be lower than this theoretical value. Therefore, a simulation-based analysis is required to allow systematic experimentation with changing load scenarios.

10.4 Model Building

Based upon the conceptual model and the results from previous analysis the simulation model is built by use of the DOSIMIS-3 simulation package which is introduced first before model structure and parameters are described.

10.4.1 The DOSIMIS-3 Simulation Package

DOSIMIS-3 (http://www.sdz.de) belongs to the class of simulation tools using building blocks for model representation. With this, model building and simulation is intended to be brought closer to the experts in an application area enabling them to implement and use a simulation model themselves [1].

DOSIMIS-3 is specialised in the answering of questions related to functionality and performance measures of logistics systems and processes. It is widely deployed in industry as well as logistics education and training in German-speaking countries. The package provides an extensive library of modules (i.e. building blocks) from the material flow and logistics world, enabling model building by a few clicks on the basis of a well-structured conceptual model. The simulation model then consists of the selected modules specified by respective sets of technical, geometrical, topological and strategic parameters which are placed in a working area and logically linked to each other by so-called nodes, i.e. directed arrows free of any further information. The tool's interactive graphical user interface provides certain functionality to support model implementation, validation, experimentation and presentation of results, enabling logistics experts with a certain understanding of simulation methodology to make use of it in standard situations.

10.4.2 Model Structure

For implementing the model of the analysed material handling system model building modules with functionality adequate to simulate the real system components were used (see Fig. 10.4). Each roller or chain conveyor is represented by an individual module of the accumulating conveyor type. The only exception to this is roller conveyor RC6, which is split into three interlinked modules according to its intersections (see Fig. 10.1) to authentically implement the pallets' movement on this conveyor.

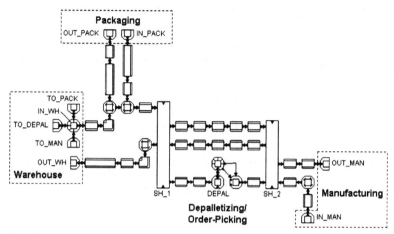

Fig. 10.4 Simulation model of the material handling system

The pallets themselves are represented by objects. They are specified by a type that encodes the object's target destination and defines its route through the system. This type code is used for flow control (see Sect. 10.4.2).

The depalletising robot is represented by a work station module (DEPAL) just consuming the total time for removing all eight pairs of boxes from the pallet and placing them on a conveyor leading to the order-picking area. As the boxes leave the studied material handling system, they are not represented in the model. Emptied pallets do only appear right after depalletising when they are forwarded to a buffer to be collected in stacks of nine pallets (STACK and B_EPAL). Completed stacks returning to the warehouse are again represented by objects.

Because intensities of the different flows are independent of each other and superpose when occupying the same resources, the model contains an individual sourcing module for each flow. In particular this means the system boundary at the warehouse where pallets for three different target destinations enter the system is represented by three independent sources (TO_Pack, TO_DEPAL, TO_MAN – see Fig. 10.4). This way of modelling allows modifying or even eliminating single flows for model validation, experimentation and analysis purposes.

10.4.3 Model Parameters

Initial parameters of model components correspond to system parameters (see Sect. 10.2.2); system boundaries operate according to the average values as used for system analysis (see Table 10.1) but applying random variations.

For manually operated system boundaries (IN/OUT_PACK, IN/OUT_MAN) normal distribution (25% deviation) represents non-permanent availability of staff. Warehouse functionality is represented in a simplified way by uniform distribution

(20% deviation) for providing outgoing pallets and by a fixed leaving time for a quasi-permanent ability to handle incoming pallets. The depalletising robot operates on a fixed-time basis as we assume a permanent and constant demand by the order-picking area and a standard loading situation of 16 boxes for all pallets.

Whereas all modules with more than one exit distribute objects according to their destination, right-of-way strategies used by modules with more than one entrance vary:

- In the warehouse (at IN_WH), flows are combined according to the FIFO rule.
- Turntable TT2 gives priority to objects provided by turntable TT1 over objects arriving from the packaging area (via RC4).
- Shuttles SH1 and SH2 always serve empty pallet stacks first and all other objects according to control strategy as explained in Sect. 10.2.4.

10.5   Verification and Validation

To verify and validate the model, i.e. to determine whether it works correctly and accurately represents the system under study, we follow a three-step approach for developing valid and credible simulation models [2]:

- The model, on the surface, needs to seem reasonable to people who are knowledgeable about the system under study. For this, we take a look at the so-called layout graphics visualising, e.g. blocking time per model component.
- The assumptions of the model need to produce appropriate results when being tested quantitatively. By use of sensitivity analysis we determine if simulation output changes significantly with random generation and termination processes.
- The simulation output data gained from the model need to closely resemble the output data that would be expected from the system under study. To test this, we compare throughput values from simulation output with those achieved from theoretical system analysis (see Table 10.1).

To reduce system complexity and get a better insight into the model's behaviour during validation, the different flows are investigated individually, i.e. we use just one of the sourcing modules to generate objects and feed them into the system. This avoids concurrent use of resources and the average hourly input at the source should directly correspond to the average output per hour at the respective sink. This method can be applied to all but one flow: empty pallets originate from the depalletising robot and not from a separate source. Therefore this flow cannot be influenced directly, but needs to be observed when investigating the flows from the warehouse or packaging to the depalletising robot. Consequently, there are five scenarios of exclusive use of resources for which we need to run simulations.

Results as shown in Table 10.2 are collected from ten repetitions of nine hours each with the first hour being dedicated to the warm-up phase and therefore excluded from statistics. To control randomness in the simulation common random numbers have been used with all flows. On this basis we are able to observe the different scenarios under similar experimental conditions and reduce the impact of randomness

Table 10.2 Parameters and results of model validation

Pallet flow	Average output (or leaving) time	Average hourly throughput Estimate	Measure	Module utilisation graphics (average hourly utilisation for exemplary 8 hour simulation run)
Warehouse – Packaging	120 s	30 pal/h	29.69 pal/h (input: 30.13 pal/h)	
Warehouse – Depalletising robot (– Warehouse)	140 s	24.28 pal/h	24.25 pal/h	
Packaging – Depalletising robot (– Warehouse)	140 s	24.28 pal/h	24.28 pal/h	
Warehouse – Manufacturing	120 s	30 pal/h	29.68 pal/h (input: 30.12 pal/h)	
Manufacturing – Warehouse	120 s	30 pal/h	30.14 pal/h (input: 30.14 pal/h)	

on the comparison of performances. For each run the hourly throughput at relevant components (i.e. the sink of each flow) is measured over eight hours which gives a total number of 80 observations per result per flow. The average value resulting from this (see Table 10.2, column 4) is compared to the maximum throughput calculated in Table 10.1 (see Table 10.2, column 3).

For all flows, the results are very close; small differences appear only with those flows linking randomly operating sources with randomly operating sinks. As the warehouse is just modelled as provider or recipient of pallets with its parameters being directly adjusted to the needs of the corresponding components, it will always be able to accept pallets or to send pallets on demand. This allows us to assume that those slight differences detected so far are not significant – a conclusion which is also confirmed by the estimated confidence intervals.

In addition to this quantitative evaluation of simulation results a qualitative model validation is carried out by use of the average hourly utilisation of model components visualised in the model layout (see Table 10.2, column 5). This graphical representation of simulation results shows which modules have been used to what extent in the course of the simulation run. Since we investigate flows individually, just those modules we expect the objects to pass should be highlighted. All other parts of the model should remain unused throughout the entire simulation run and appear much brighter in the layout. From comparing these graphical (see Table 10.2, column 5) results directly provided by the simulation package with the specification of principle pallet flows as shown in Fig. 10.2 we clearly see another match confirming the correctness and validity of the model.

To determine the model's sensitivity with regard to stochastic variation of generation and termination processes we compare the average length of the queue inside a sourcing component (i.e. the average number of objects that cannot yet enter the system) and the average blocking time of key components (i.e. components that might be critical ones for a specific flow). As shown in Fig. 10.5 varying random numbers generally influence these measures, but the observed impacts significantly differ between different flows.

For pallet flows from the warehouse to packaging and manufacturing, turntables TT1 and TT2 respectively were considered the critical elements. Their blocking time varies between 0 and 60% (TT1 – see Fig. 10.5a) and 0 and 40% (TT2 – see Fig. 10.5b). Since the overall average blocking times are pretty low (18% for warehouse – packaging and 8% for warehouse – manufacturing), especially the high values can be considered exceptions. This assumption is also supported by the fact that with both flows hardly any pallet jams could be observed towards the respective sourcing components: at maximum just one or two pallets were waiting inside TO_PACK or TO_MAN to enter the system. In the further course of a flow the number of queuing pallets and the blocking time of a component increases as shown for RC6 in Fig. 10.5b. Here, not just the average percentage of blocking time per simulation run is always significantly higher than with the source TO_MAN, but also the average blocking time over all observations (38%). This clearly confirms the sink as initiator of pallet jams and limiting component.

The pallet flow from packaging to the depalletising robot (see Fig. 10.5c) also shows widely ranging blocking times at TT2 (40 to 80%) and even further not a single simulation run without any blocking of TT2. The average percentage of blocking time is also pretty high (about 65%) indicating a significant system overload. As a result at least temporary pallet jams back to the respective source IN_PACK cannot be avoided. The same effect is even more visible with the pallet flow from the warehouse to the depalletising robot (see Fig. 10.5d). Here, the average block-

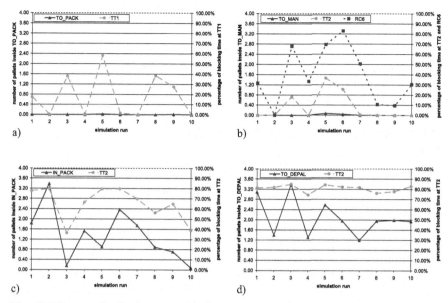

Fig. 10.5 Impact of pallet jams on critical components. **a** Warehouse – Packaging. **b** Warehouse – Manufacturing. **c** Packaging – Depalletising robot. **d** Warehouse – Depalletising robot

ing time of TT2 always reaches values between 75 and 85%; on average more than two pallets queue up in the source TO_DEPAL.

Whereas blocking times vary a great deal with random generation and termination processes, the average hourly throughput at sources and sinks show just minor differences of ±1%. From this we can conclude that flow-specific throughputs are not directly influenced by stochastic processes and 10 replications of 8-hour simulation runs should be sufficient for the intended performance investigations; the model can be considered correct and valid beyond any doubt.

10.6 Simulation Experiments

10.6.1 Objectives and Procedure

The load limit of the material handling system can be measured by its performance in providing manufacturing, packaging and order-picking areas (i.e. the depalletising robot) with pallets or removing pallets from these areas. It is represented by the total throughput of the concurrent flows if flow-specific input and output volumes are well-balanced even in short observation periods. As the load limit of the entire system results from superposing concurrent pallet flows, the maximum total of all parallel throughputs – measured at OUT_PACK, OUT_MAN, DEPAL and

OUT_WH – can be used as performance indicator. Additionally, further components like SH_1, SH_2, TT1 and TT2 are observed, because they are used by different flows and eventually might form conflict points. Because of this, we do not measure throughput at these points only, but also collect data on the percentage of blocking times and on the length of queues in the run-up to these points.

Experimentation strategy consists of two stages. First, all flows are investigated separately to determine flow-specific performance limits with exclusive use of resources. For each flow the performance is limited by the termination process of the respective sink. The generation and output behaviour of the source has to be adjusted to this in such way that no pallet jams appear. This is to be achieved by reducing the performance of a source in steps until all pallets put into the system run through it without any waiting time (or component blocking respectively). The total of these flow-specific performance limits forms the maximum load that theoretically could be mastered by the system and serves as an evaluation basis for analysing concurrent flows. The latter is subject to the second stage of experimentation with variation of superposed flows to understand to what extent concurrent use of resources reduces system performance. From superposing all flows according to their individual limits as determined before, we expect component blocking and pallet jams; critical elements used by different flows might form bottlenecks of the system. Starting from the maximum loads we therefore proportionally reduce intensity of all flows step-by-step until all bottlenecks (outside the sinks) have disappeared.

Finally, we pay special attention to the two flows that might split towards two sinks (see Fig. 10.2), i.e., manufacturing – warehouse (or depalletising robot) and packaging – depalletising robot (or warehouse). Keeping the total number of pallets forwarded to the depalletising robot at constant level we vary the percentage of pallets provided by different sources (see Table 10.3). The aim of these experiments

Table 10.3 Plan of experiments for splitting flows

No.	IN_PACK – DEPAL vs. IN_PACK – OUT_WH	No. of pallets IN_PACK – DEPAL	Missing pallets at DEPAL	Output time TO_DEPAL (uniform)
1	80% : 20%	24 pal/h	24.25 – 24 = 0.25 pal/h	No generation
2	50% : 50%	15 pal/h	24.25 – 15 = 9.25 pal/h	[312.3, 468.5] s
3	20% : 80%	6 pal/h	24.25 – 6 = 18.25 pal/h	[158.3, 237.5] s
4	0% : 100%	0 pal/h	24.25 – 0 = 24.25 pal/h	[119.1, 178.7] s

No.	IN_MAN – DEPAL vs. IN_MAN – OUT_WH	No. of pallets IN_MAN – DEPAL	Missing pallets at DEPAL	Output time TO_DEPAL (uniform)
5	80% : 20%	24 pal/h	24.25 – 24 = 0.25 pal/h	No generation
6	50% : 50%	15 pal/h	24.25 – 15 = 9.25 pal/h	[312.3, 468.5] s
7	20% : 80%	6 pal/h	24.25 – 6 = 18.25 pal/h	[158.3, 237.5] s
8	0% : 100%	0 pal/h	24.25 – 0 = 24.25 pal/h	[119.1, 178.7] s

is to investigate how utilisation of particular components change and in which way the system's performance limit is influenced.

10.6.2 Simulation Results

10.6.2.1 Investigation of Separate Flows

To determine flow-specific performance limits we based simulation experiments on those stochastic processes which caused the highest blocking time percentage in the validation runs (see Fig. 10.5). Starting from sourcing parameters of the validation runs we reduced the system load for each flow until nearly no blocking was observed (for an example see Fig. 10.6). Here, blocking times of about 2% or even less were considered insignificant as they represent less than 10 minutes of the total observation period of 8 hours. The flow-specific load the system could master without blocking was used as estimate of the respective performance limit. As it was determined by looking at one particular representation of stochastic processes only, it necessarily had to be verified by nine further simulation runs with varying random numbers.

As comparison of flow-specific results shows, sourcing performance had to be reduced to varying extent (see Table 10.4). Whereas the warehouse-packaging flow was nearly free of blocking with a 97% performance level of the source, performance of the source of the pallet flow from packaging to the depalletising robot had to be reduced to 90% of the average demand of the respective sink to cut down blocking times to the intended level. Even though, additional stochastic influences might be taken into consideration with the latter flow these differences remain significant and

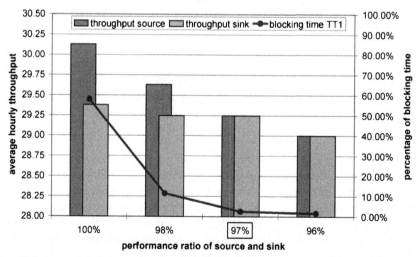

Fig. 10.6 Determining the performance limit of the warehouse-packaging flow

Table 10.4 Parameters of flow-specific performance limits

Pallet flow	Sourcing parameters		Average hourly throughput		Blocking time
	Perfor-mance	Output time	Source	Sink	
Warehouse – Packaging	97%	[99.0, 148.4] s (uniform)	29.21 pal/h	29.17 pal/h	TT1: 0.67%
Warehouse – Depalletising robot	94%	[119.1, 178.7] s (uniform)	24.29 pal/h	24.13 pal/h (return flow: 2.69 stacks/h)	TT2: 1.54%
Packaging – Depalletising robot	90%	155.6 ± 38.9 s (normal)	23.03 pal/h	22.98 pal/h (return flow: 2.57 stacks/h)	TT2: 1.90%
Warehouse – Manufacturing	95%	[101.0, 151.6] s (uniform)	28.67 pal/h	28.60 pal/h	RC6: 0.64%
Manufacturing – Warehouse	100%	120 ± 30 s (normal)	30.14 pal/h	30.14 pal/h	No

need to be questioned. One possible explanation might be that the warehouse–packaging flow just touches the material handling system passing by one more complex element (TT1) only. In the case of this particular flow this component is used as a pure conveying element without any additional functionality. Furthermore, the second flow towards the depalletising robot (from the packaging area) also required a significant reduction of the sourcing performance (down to 94%), which points to the return flow of empty pallets as a reason for an increased load with shuttle SH1 and consequently a higher chance of delaying pallets at TT2.

The need to reduce sourcing performances also influences flow-specific load limits expressed by the average hourly throughput at the respective sinks. The maximum possible workload of the entire system now results from summing up throughputs according to flow-specific performance limits from Table 10.4. As our investigation was based on separate flows, but the depalletising robot receives pallets from two sources (packaging and warehouse) at the same time, each of both flows (and the return flows of empty pallet stacks to the warehouse as well) is assumed to deliver half of the pallets needed (objects = pallets or stacks):

$$29.17 + \frac{24.13 + 22.98}{2} + \frac{2.69 + 2.57}{2} + 28.60 + 30.14 = 114.09 \text{ obj/h} .$$

This value is a little less than the theoretical performance limit as estimated in Sect. 10.3 because of random behaviour of sources and sinks. Nevertheless it is still based on the simplification of separate flows and exclusive use of resources. Therefore, in the next part of the experiments we need to investigate what happens when all flows pass the material handling system at the same time concurrently using resources. From this, we clearly expect increased blocking times which finally might lead to reduced flow intensities and decrease system throughput.

10.6.2.2 Analysis of Superposed Flows

For superposing flows we basically use sourcing parameters according to flow-specific performance limits (see Table 10.4). Again we must pay special attention to the flows towards the depalletising robot and reduce output performance of the respective sources to the half (i.e. doubling output times). Based upon this we run ten simulation runs with varying random numbers (observation period 480 minutes) to achieve the average hourly throughput of all sources and sinks.

Analysing the results (see Table 10.5) we consider surprisingly and in contrast to our expectations that the material handling system actually can cope with the maximum flows from the exclusive-use-of-resources scenario. Total system throughput and flow-specific throughputs differ just slightly from previous results although certain components show drastically increased blocking times (see Fig. 10.7). One possible reason for this might be seen in the design of flow-specific loads as this was targeted to nearly eliminate blocking times when resources are used exclusively (see Sect. 10.6.2.1). This seems to have provided critical components with sufficient flexibility with regard to concurrent flows to avoid permanent, through-put-reducing pallet jams. On the other hand, the significant increase of blocking times in Fig. 10.7 demonstrates that this system-immanent reserve performance is hence limited and an even the slightest load increase might decrease system performance.

To validate this assumption, further experiments have been carried out in which the output of all sources has proportionally been increased. As random numbers were used that previously have produced the largest amount of blocking times (see Fig. 10.7, simulation run 4), just one simulation run per experiment was sufficient to see effects and derive conclusions about the system's performance (see Table 10.6):

Table 10.5 Input parameters and results from simulation runs with concurrent flows

Pallet flow	Input parameters		Average hourly throughput	
	Source	Sink	Expected	Measured
Warehouse – Packaging (97%)	[99.0, 148.4] s (uniform)	120 ± 30 s (normal)	29.17 pal/h	29.03 pal/h
Warehouse – Depalletising robot (94%)	[238.2, 357.4] s (uniform)	140 s (fixed)	12.07 pal/h	12.14 pal/h
Packaging – Depalletising robot (90%)	311.2 ± 77.8 s (normal)	140 s (fixed)	11.49 pal/h	11.42 pal/h
Warehouse – Manufacturing (95%)	[101.0, 151.6] s (uniform)	120 ± 30 s (normal)	28.60 pal/h	28.41 pal/h
Manufacturing – Warehouse (100%)	120 ± 30 s (normal)	0.1 s (fixed)	30.14 pal/h	30.23 pal/h
Depalletising robot – Warehouse	No sourcing	0.1 s (fixed)	2.63 stacks/h	2.59 stacks/h
System total			114.09 obj/h	113.82 obj/h

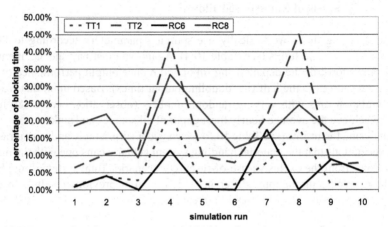

Fig. 10.7 Impact of randomness on average blocking times (concurrent flows)

whereas total throughput of the system and utilisation of shuttles remain at nearly the same level, blocking times with all critical components individually and on average significantly increase. Although more pallets are in the system the number of pallets (or stacks of empty pallets) received at the sinks does not change. Consequently, any further increase of the load to be mastered by the system is not useful and the system's performance limit remains at the level of 114.09 objects per hour as already achieved for an exclusive use of resources.

Since an increase of the number of pallets being fed into the system does not improve system performance, we investigate the influence by splitting flows according to the plans from Sect. 10.6.1. Comparing the outcomes of these experiments with the constellation of superposed 'standard' flows (i.e. without splitting) we can clearly determine significant changes in terms of both the throughput (see Fig. 10.8a) and the blocking or utilisation situation. This effect is a logical consequence of the experimental design. The demand for pallets with the depalletising robot remains stable independent of the source of the pallets. If – as we do now – not all pallets from the packaging are forwarded to the depalletising robot, but to a growing extent are sent to the warehouse instead, the missing amount of pallets has to be compensated by additional deliveries from the warehouse. The total number of pallets to be mastered by the material handling system increases by the same amount. As the flows in the 'standard' version were already adjusted to the system's performance limit, these extra pallets again overload the system. Whereas the flows towards packaging and manufacturing do not show any difference, the flows towards the depalletising robot and the warehouse do not reach the performance level as before – no matter which splitting ratio has been used. The main reason for this is the almost 100% utilisation of shuttle SH1 causing a significant increase in blocking times at RC8 and TT2 (and further on TT1 as well).

Splitting the flow from manufacturing towards the depalletising robot and the warehouse and at the same time sending all pallets from packaging to the warehouse and no longer to the depalletising robot affects all other pallet flows. In con-

Table 10.6 Simulation results for increased sourcing performance (basis: simulation run no. 4)

		Sourcing performance as before	1% increase of sourcing performance	2% increase of sourcing performance
Throughput	OUT_PACK	28.75 pal/h	29.00 pal/h	29.25 pal/h
	DEPAL (from packaging)	11.63 pal/h	11.75 pal/h	11.75 pal/h
	DEPAL (from warehouse)	12.38 pal/h	12.50 pal/h	12.63 pal/h
	OUT_MAN	28.50 pal/h	28.75 pal/h	29.00 pal/h
	OUT_WH (from manufacturing)	31.00 pal/h	31.00 pal/h	30.13 pal/h
	OUT_WH (from DEPAL)	2.63 stacks/h	2.63 stacks/h	2.63 stacks/h
	System total	114.89 obj/h	115.63 obj/h	115.39 obj/h
Blocking time	TT1	22.14%	24.40%	39.39%
	TT2	42.81%	52.80%	68.42%
	RC6	11.17%	24.99%	15.88%
	RC8	33.38%	46.25%	59.52%
	Average	27.38%	37.11%	45.80%
Utilisation	SH1	92.89%	93.94%	94.17%
	SH2	62.57%	63.04%	62.94%
	Average	77.73%	78.49%	78.56%

trast to splitting the packaging flow we can even observe a reduced utilisation of shuttle SH1, although pallets queue even further. This appearance results from situations when roller conveyors in the run-up to the depalletising robot are fully occupied by pallets and shuttle SH1 cannot deliver further pallets to this destination for the moment. If only pallets which also need to be moved towards the depalletising robot are waiting for transportation by shuttle SH1, we have a temporary deadlock situation at SH1 causing waiting times with the shuttle.

In the end any splitting of flows leads to a system overload and more or less significantly reduces system performance (see Fig. 10.8a). Here, SH1 and TT2 are the main bottlenecks which cannot be removed by any modification of the source or sink parameters. Because of this, the system's maximum throughput and performance limit still remains at the level of 114.09 pal/h as achieved from investigations with exclusive use of resources.

10.7 Conclusions

Because of the various crossing pallet flows and the conflict potential resulting from this, it was not an easy task to characterise the performance limit of the material

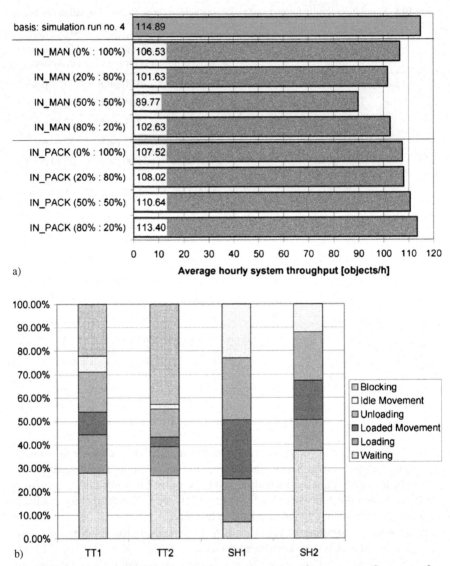

Fig. 10.8 Analysis of simulation results for concurrent use of resources. **a** System performance for splitting flows. **b** Working states of critical components

handling system under investigation. The performance limit was defined as the pallet throughput that is achieved throughout the system without any pallet jam with the sources or any defined bottleneck elements. From the beginning of the study the possibility to experimentally increase intensity of flows was limited to manually operated transfer points into the packaging and manufacturing areas and the continuous demand with the depalletising robot already defined the borderlines.

First experiments dealt with separate flows exclusively using the resources. From this idealisation we gained maximum values forming the basis for further investigation of concurrent flows. Surprisingly, superposition of flows at their individual limits did not produce the conflicts, bottleneck situations and pallet jams expected. Therefore, the performance limit of the system as a whole is equal to the sum of the maximum throughput of the separate flows. The main reason for this was the limitation by the manual sinks and the depalletising robot, whereas the conveying components of the material handling system would have allowed a higher performance level. Instead those components were not used at their limits and were therefore able to master higher loads from concurrent flows. This is confirmed by an analysis of the working states of the most critical conveying components, turntables TT1 and TT2 as well as shuttles SH1 and SH2 (see Fig. 10.8b). Even enormous blocking times of up to 40% with the turntables did not hinder the system in coping with the total load, as at the same time also (idle) waiting times of more than 25% were to be observed. Further experiments to split flows did always end in reduced system performance.

The project presented in this chapter aimed to analyse a small system hosting some process complexity because of the concurrent flows. From a logistical point of view system design was not optimal, but as investigations have shown the components were well balanced and matching current requirements. Any increase of the performance beyond the current limit of 114.09 pallets or stacks per hour would only become possible if system design and operation is changed and concurrent use of resources is reduced.

10.8 Questions

1. What is to be taken into consideration when determining the performance limit of a material handling system?
2. What is the difference between a system's performance limit and its technical capability?
3. For which kind of material flow components do rules for process control need to be defined and why?
4. What is forward control and what does it mean for object movement in a material flow?
5. Why it is useful to apply the push principle when estimating a system's performance limit?
6. How can the performance limit be estimated for a system with concurrent flows?
7. What are suitable methods to validate a material flow simulation model?
8. What is a system-immanent reserve performance and how does it influence determination of a system's load limit?

References

[1] Page B, Kreutzer W (2005) The Java simulation handbook – simulating discrete-event systems in UML and Java. Shaker, Aachen
[2] Law A (2007) Simulation modeling and analysis. McGraw-Hill, New York

Chapter 11
Vessel Traffic in the Strait of Istanbul

Ö. S. Ulusçu, B. Özbaş, T. Altiok, İ. Or and A. Ö. Almaz

Abstract This study develops a high-fidelity simulation model for the vessel traffic in the Strait of Istanbul using the Arena simulation tool. The simulation model is developed mainly for risk analysis and mitigation purposes. In addition, it is utilised to perform scenario analysis to investigate the effect of various system variables such as arrival frequency, number of pilots and number of tugboats on the system performance. The model incorporates an algorithm to schedule vessel entrances to the Strait using the decisions made by the Vessel Traffic Services Authorities. In this regard, the Strait Traffic Rules and Regulations, transit vessel profiles along with local traffic and other vessels, pilotage and tugboat services, meteorological and geographical conditions are all considered in the model to provide a tool to analyse the effectiveness of various current and future policies and decisions regarding management of traffic, risks and vessel delays in the Strait of Istanbul.

11.1 Introduction

The Strait of Istanbul has for centuries been one of the world's most strategic waterways. As the Black Sea's sole maritime link to the Mediterranean and the open ocean beyond, it remains a vital passageway not just for trade but for the projection of military and political powers. Today, this narrow passage runs through the

Özgecan S. Ulusçu, Tayfur Altiok and Alper Ö. Almaz
Rutgers University, USA
ozgecanu@eden.rutgers.edu, altiok@rci.rutgers.edu, alperalmaz@hotmail.com

Birnur Özbaş and İlhan Or
Boğaziçi University, Turkey
birnur@ozbas.com.tr, or@boun.edu.tr

Y. Merkuryev et al. (eds.), *Simulation-Based Case Studies in Logistics*
© Springer 2009

heart of Istanbul, home to over 12 million people and some of the world's most celebrated ancient sites.

The Strait of Istanbul is approximately 31 km long, with an average width of 1.5 km. At its narrowest point between Kandilli and Bebek, it measures a mere 698 m. It takes several sharp turns, forcing the ships to alter course at least 12 times, sometimes executing turns of up to 80 degrees. Navigation is particularly difficult at the narrowest point, as the vessels approaching from opposite directions cannot see each other around the treacherous bends.

In addition to its winding contour, the unpredictable countervailing currents that may reach 7 knots pose significant danger to ships. Surface currents in the Strait flow from the Black Sea to the Sea of Marmara, while submarine currents 50 feet below the surface run in the opposite direction. Within bays and near point bars, these opposing currents lead to turbulence. The unpredictable climate brings about further danger. During storms with strong southerly winds, the surface currents weaken or reverse in some places, making it even harder to navigate. Not surprisingly, all these elements can easily cause vessels transiting the Strait to veer off course, run aground or collide. An aerial view of the Strait is given in Fig. 11.1.

The current international legal regime governing the passage of vessels through the Turkish Straits (Istanbul and Çanakkale) is the 1936 Montreux Convention. Although this instrument provides full authority over the straits to the Turkish government, it asserts that in time of peace, merchant vessels are free to navigate the straits without any formalities. When the Convention was put in place, less than 5,000 vessels used to pass through the Strait of Istanbul annually. Today, the changes in the shipping and navigational circumstances have led to an eleven-fold increase in the maritime traffic through the Strait.

Several reasons contributed to this immense increase in traffic. The Turkish Straits provide the only maritime link between the Black Sea riparian states and the Mediterranean, forcing the states to rely on the straits for foreign trade. The opening of the Main–Danube canal has linked the Rhine to the Danube, linking the

Fig. 11.1 The Strait of Istanbul (Courtesy of Google Inc.)

North Sea and Black Sea. Traffic originating from the Volga–Baltic and Volga–Don waterways has also increased in recent years. Still, the most alarming increase in traffic is observed in the number of vessels carrying dangerous cargoes. The fall of the Soviet Union in 1991 has led to the emergence of newly independent energy-rich states along the Caspian Sea. Currently, oil and gas from Azerbaijan, Turkmenistan and Kazakhstan reach the western markets through the Turkish Straits [1].

In order to ensure the safety of navigation, life and property, and to protect the environment, the Turkish government adopted unilaterally the 1994 Maritime Traffic Regulations for the Turkish Straits and Marmara Region [2]. In 1998, the rules were revised and the 1998 Reviewed Regulations were adopted. These regulations include extensive provisions for facilitating safe navigation through the straits in order to minimise the likelihood of accidents and pollution.

The complexity of the operation at the Strait of Istanbul motivates us to study all aspects of the traffic. Our goal is to develop tools to assist in day-to-day vessel scheduling as well as produce operational policies reducing congestion in the passage. This will help in determining threshold levels on accepting an arriving vessel into the Strait based on the day's cargo and vessel schedule. We have developed a simulation model to study the transit vessel traffic in the Strait of Istanbul. A scenario analysis is performed in order to evaluate the impact of several parameters on the system performance.

The following study involves the risk analysis of the dangerous cargo vessel traffic in the Strait by superimposing a mathematical accident risk model over the simulation model. The mathematical risk model is based on probabilistic arguments regarding instigators, situations, accidents, consequences and historical data as well as subject-matter expert opinions. Risks are calculated in the simulation model with respect to surrounding geographical, meteorological and traffic conditions.

11.2   Vessel Traffic in the Strait of Istanbul

Each year, more than 50,000 transit vessels pass through the Strait of Istanbul, carrying various cargoes ranging from dry goods to petroleum products as shown in Fig. 11.2. After arriving at the entrances, the vessels may anchor for various reasons including health inspection, loading food or refuelling. All vessels, anchored or not, wait in the queue until they are allowed to transit. The Strait is divided into two traffic lanes. The vessels are permitted to enter the Strait one at a time from each entrance. The vessel traffic may be interrupted due to poor visibility, high currents, and other factors such as lane closures caused by vessel accidents or sporting events. Vessels do not stop in the Strait since they may create a high-risk situation for other vessels and the environment. The key locations in the Strait of Istanbul are given in Fig. 11.3.

Fig. 11.2 Fatih Sultan Mehmet Bridge – the narrowest part of the Strait

Fig. 11.3 Key locations in the Strait of Istanbul (courtesy of the Turkish Navy Office of Navigation, Hydrography and Oceanography)

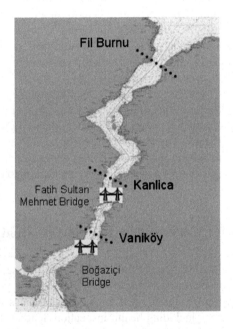

The increases in the traffic between 2003 and 2005 are 17% and 28%, respectively. A total of 54,623 vessels passed through the Strait in 2005. According to the data obtained from the VTS (Vehicle Tracking System) Office of the Strait of Istanbul, 19% of the transit vessels carry dangerous cargo such as natural gas, chemicals, oil, nuclear waste and derivatives through the Strait. The US Energy Information Administration estimated that 3.1 million barrels of oil pass through the Strait every day in 2004 [3]. On the other hand, 79% of the vessels are general cargo vessels while passenger vessels constitute the rest, 2%.

11.3 Regulations

Vessels approaching the Strait are required to provide sailing plans prior to their passage. They are required to submit their sailing plans (SP 1) to VTS to advise their arrival details along with vessel and cargo information, and tugboat/pilot request status at least 24 hours prior to their arrival. In addition to SP 1, vessels are required to submit another report called Sailing Plan 2 (SP 2), which provides further details to the authorities at least 2 hours prior to their arrival.

The passage through the Strait is regulated using the following rules:

- Passage of vessels longer than 200 m is restricted to daytime.
- Navigational speed limit at the Strait is 10 nautical miles per hour, unless a higher speed is needed to maintain good steerage.
- Transiting vessels must maintain a minimum following distance of 8 cables (≈ 1 mile) while passing through the Strait.
- The passage of vessels carrying dangerous or hazardous cargo is regulated under Reg. 25 letter d, which states that when a northbound (southbound) vessel (>150 m) carrying dangerous cargo enters the Strait, no northbound (southbound) vessel with the same characteristics is permitted until the former vessel reaches the Fil Burnu (Boğaziçi Bridge).
- Non-stopover vessels may anchor for up to 48 hours without a clean Bill of Health.
- Overtaking is forbidden unless absolutely necessary. It is not allowed between the Vaniköy and Kanlıca points under any circumstances.
- Traffic in the Strait may temporarily be suspended in the case of force majeure situations, collision, grounding, fire, public security, pollution and similar occasions, surface or underwater construction, and the existence of navigational dangers.
- Incoming traffic from the opposite direction is suspended when a vessel with a length of 250–300 m passes through the Strait.
- Traffic is suspended in both directions when a vessel greater than 300 m in length transits the Strait.
- Vessels may not be allowed to the Strait based on the surface currents and their speed.
- Passage of vessels may be restricted under certain visibility conditions to ensure the safe navigation.

11.4 Literature on Modelling of Waterways

The research conducted on modelling and performance analysis of narrow waterways is scarce. Some of the work published on the topic in addition to studies focusing on the Panama and Suez Canals are discussed below.

A simulation model of the transit traffic in the Strait of Istanbul is presented in [4]. Specifically, the focus is on the variation of waiting times resulting from different transit vessel arrival frequencies. The results of the simulation model, and the effects of probable increases in marine traffic due to new oil pipelines, are discussed.

In [5] a simulation model is proposed to estimate the number of vessel interactions in the current system and their increases caused by three alternative expansion plans in the San Francisco Bay. The increase in the number of situations where ferries are exposed to adverse conditions is evaluated by comparing the outputs.

The simulation study of the transit maritime traffic in the Strait of Istanbul presented in [6], focuses on the modelling of the entrance procedures based on vessel types and lengths, prioritisation of vessels, pilotage and tugboat services. This model incorporates an older application of rules and regulations for vessel entrance.

Finally, in [7, 8] a functional simulation model of the maritime transit traffic in the Istanbul channel is presented. The objective is to perform scenario analysis to analyse the effectiveness of various operational policies.

11.5   Modelling the Transit Vessel Traffic

We have developed a high-fidelity simulation model representing the vessel traffic within the Strait of Istanbul using the Arena simulation software (http://www.rockwellautomation.com/rockwellsoftware). The simulation model is developed to investigate the effect of various system attributes such as arrival frequency, number of pilots and number of tugboats on the system performance as well as testing scheduling policies.

Due to various reasons, including the Marmaray tunnel construction project, the transit traffic may be single-directional or bi-directional. For instance, during the daytime, the transit traffic is allowed in only one direction, while the night-time traffic has been bi-directional in the years 2005 and 2006. This is what we have assumed in this study.

11.5.1   Vessel Arrivals

Vessels are created based on the cargo type and vessel length categories. In other words, every combination of vessel length and cargo type is generated using a unique arrival process deciphered from data. Although the logic for traffic scheduling of both directions is the same, the inter-arrival time distributions are different for northbound and southbound vessels since they are coming from different ports and seas. Entities representing different types of vessels are generated according to SP 2 submitted by approaching vessels.

Upon arrival, all vessels are assigned the following attributes: vessel length, vessel class, speed, tugboat request indicator, pilot request indicator, anchorage indicator, anchorage duration, and stopover indicator.

The distributions used for the above attributes are unique to each type of vessel of a certain length. Indicator values for tugboat request, pilot request, anchorage and stopover are assigned according to the corresponding percentages. Following its creation, a vessel entity is sent to the anchorage area if its *Anchorage Indicator* equals 1, and waits for its *Anchorage Duration*. After it leaves the anchorage area the entity joins the queue of its *Vessel Class*. The vessel classes created for scheduling purposes are shown in Fig. 11.4. Each vessel waits in its queue until one of the entities representing daytime or night-time schedulers removes it from the queue.

Class T6 and Class A vessels have priority over any other vessel since they can only pass through the Strait during daytime. In addition, Class T6 vessels have priority over Class A vessels since their passage is subject to special permissions from the authorities and it requires two-way traffic suspension. Vessels have the following priority structure for scheduling purposes:

$$T6 > A > B > C > P > E > D \qquad (11.1)$$

11.5.2 Resources

When a vessel is removed from its class queue, it seizes the necessary resources to enter the Strait based on its indicator attributes. These resources are pilots and tugboats, which are grouped into two categories: northbound and southbound. If there are no resources available, a vessel requesting resources is not removed from the queue. Once a vessel seizes all resources it needs, it enters the Strait.

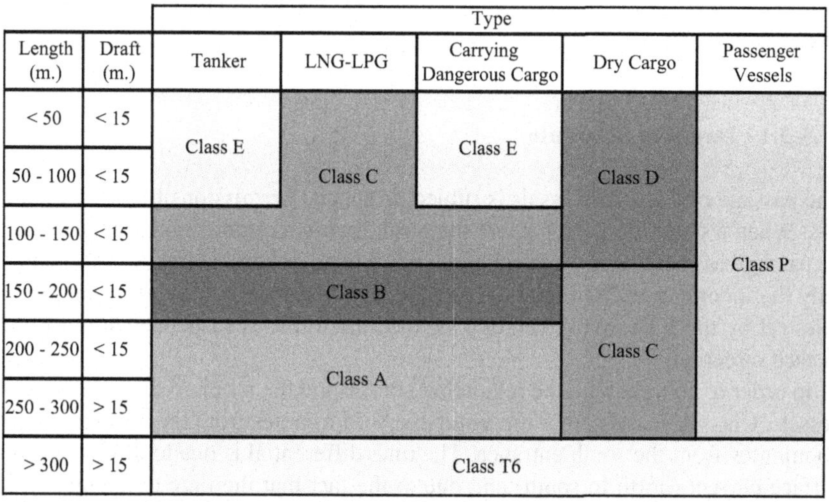

Length (m.)	Draft (m.)	Type				
		Tanker	LNG-LPG	Carrying Dangerous Cargo	Dry Cargo	Passenger Vessels
< 50	< 15	Class E		Class E		
50 - 100	< 15		Class C		Class D	
100 - 150	< 15					
150 - 200	< 15		Class B			Class P
200 - 250	< 15				Class C	
250 - 300	> 15		Class A			
> 300	> 15			Class T6		

Fig. 11.4 Vessel classes for scheduling vessel entrances into the Strait

The seized resources are released by the vessel when it completes its passage. The released resources are designated to be available in the direction of the entrance they are released at. Due to various reasons, including the Marmaray tunnel construction project, the transit traffic may be single-directional or bi-directional. For instance, during the daytime, the transit traffic is allowed in only one direction while the night-time traffic was bi-directional in the years of 2005 and 2006. This is what we have assumed in this study. Current practice is that pilots are taxied back to the open entrance every time the number of available pilots on the opposite direction reaches five.

On the other hand, when a pilot is released at an entrance by a vessel during a night-time passage, the number of available pilots in that entrance is checked. If this number is greater than half of the total pilot capacity and the difference is at least five, then the released pilot and four others are taxied back to the other entrance. Further, tugboats return one by one if there are less than two tugboats available in the open entrance and more than two in the other end.

11.5.3   Vessel Scheduling

Turkish Straits VTS schedules the vessels entering the Strait based on their waiting times, and priorities. In addition, the regulations in place, the number of vessels in both directions and the number of available pilots play a role in the scheduling decisions. We have developed a mathematical model of the current scheduling practice at VTS and incorporated it into the vessel traffic simulation model. Its fundamental philosophy is to schedule the vessels with highest waiting time first while giving priority to large vessels carrying dangerous cargo.

Vessels of Class T6 and Class A may pass through the Strait only during daytime. Therefore, different scheduling policies are used for daytime and night-time vessel traffic.

11.5.3.1 Daytime Schedule

The passage of Class T6 vessels is subject to special permissions from the authorities. When a Class T6 vessel enters the Strait, two-way traffic is suspended during its passage as mentioned earlier. On the other hand, when a Class A vessel enters, only the incoming traffic (opposite direction) is suspended. Also, according to the rules set by the VTS, in a given day the daytime traffic is suspended at most once in each direction.

In order to comply with the regulation concerning the required distance between vessels, Class A vessels enter the Strait every 75 minutes from the north and every 90 minutes from the south entrance. The time differential is due to direction of the surface current (north to south) and due to the fact that the time to navigate from the south entrance to Fil Burnu is greater than the one from the north entrance to Boğaziçi Bridge. However, Class C vessels may follow each other at 30-minute intervals. Furthermore, Class D, E and P vessels may enter every 10 minutes. A typical order of vessels entering the Strait during daytime is given in Figs. 11.5 and 11.6. The difference in the northbound and southbound schedules is due to the aforementioned differences in inter-entrance times to the Strait.

Since Class T6 and Class A vessels can only pass through the Strait during daytime, the total number of these vessels passing in a day is limited by the daytime duration. This duration is seasonal and changes throughout the year. Also, VTS gives Class T6 and Class A vessels priority for daytime scheduling. Every morning, two hours before sunrise, VTS operators determine the daytime transit vessel schedule for that day, considering the following policies:

- Daytime starts at dawn and ends at sunset.
- Vessels with highest waiting times are scheduled to enter the Strait first.
- Class T6 vessels have priority over Class A vessels.
- Stopover vessels have higher priority than non-stopover vessels.
- Southbound stopover vessels have higher priority than northbound stopover vessels.

They first decide on the Class T6 and Class A vessels that will pass that day in both directions based on the list of vessels that have submitted their SP 2 but have not yet entered the Strait. According to the scheduling algorithm, the Class T6 and Class A vessels are first sorted in decreasing order of *adjusted waiting times* within their respective classes. The two sorted vessel groups are then combined in a *tentative*

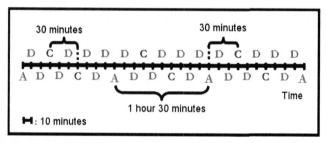

Fig. 11.5 Typical order of northbound vessels entering the Strait

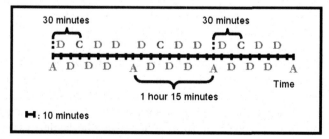

Fig. 11.6 Typical order of southbound vessels entering the Strait

list, in which Class T6 vessels precede Class A vessels. This list includes all vessels ready to enter the Strait from both directions.

Next, the number of Class T6 and Class A vessels in the *tentative list* that will be able to enter the Strait that day is determined considering the *start time*, and the *maximum operational duration* of the daytime schedule. The new list forms the *initial schedule* of Class T6 and Class A vessels that will enter the Strait that day from both directions. The initial schedule is then compared with two other scenarios, where one more northbound and southbound Class T6 and Class A vessel is scheduled, respectively while the total number of vessels in the schedule remains the same. The objective is to schedule vessels with higher adjusted waiting times first and to schedule more Class T6 and Class A vessels from the entrance, which has more vessels waiting. Also, there should be a sufficient number of Class P, E and D vessels to schedule in between consecutive Class A vessels. Therefore, in order to compare the scenarios, a scoring objective function is evaluated for each. The scenario with the highest objective value is chosen as the *final schedule*.

Then, they schedule the rest of the vessels to enter in between consecutive Class A vessels according to order plans depicted in Figs. 11.5 and 11.6 based on the priority $P > E > D$. Since two-way traffic is suspended during the passage of Class T6 vessels, no other vessel is scheduled in between two consecutive Class T6 vessels or between a Class T6 and a Class A vessel.

Finally, the initial direction of the daytime schedule is selected. The objective is to start the daytime schedule in the direction with the greater number of vessels waiting, highest total waiting time of vessels, and greater number of Class T6 and Class A vessels scheduled in the chosen scenario. In order to compare the vessel congestion in two entrances, a score value is assigned to each direction and the one with higher score is selected.

The daytime schedule described above is the initial schedule determined two hours before the entrance opens. Additionally, at the end of daytime traffic in each direction, the schedule is updated if there is any new Class A vessel. If the *maximum operational duration* is not totally utilised by the initial schedule, the new Class A vessel is added at the end.

11.5.3.2 Night-Time Schedule

In contrast with the daytime traffic, there is a two-way traffic flow during night-time. Class B vessels are the most critical vessels in terms of their type of cargo and length among all the vessels that can pass through the Strait at nights. These vessels may enter the Strait at 60-minute intervals. A typical order of vessels entering the Strait during night-time is given in Fig. 11.7. Regulations state that while a Class B or C vessel navigates through the Strait at night, no Class B, C or E vessel is allowed to enter in the opposite direction. Each night, depending on the congestion at the entrances, the passage of the aforementioned classes is allowed first in one direction and then the other. This procedure is carried out once in a given night. Thus, in order to schedule the night-time traffic, we use the daytime vessel scheduling algorithm by replacing Class A vessels with Class B vessels.

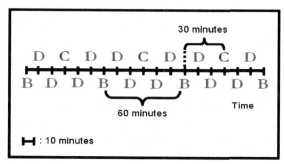

Fig. 11.7 A typical order of vessels entering the Strait during night-time

11.5.4 Lane Structure

The transit maritime traffic in the Strait is regulated within officially established traffic lanes. In the model, the pre-determined vessel routes are arranged to coincide with the centre lines of the official lanes. In the simulation model, the Strait of Istanbul is divided into 22 slices (each 8 cables long) to monitor the traffic and specifically the overtaking situations. The vessels start their travel at the entrance station of a slice and navigate continuously until the next slice is reached.

Overtaking is permitted in the Strait except at the narrowest part between Kanlica and Vaniköy points. A vessel passes another vessel travelling in front of it if it reaches the next station before the latter does. There can be no more than four vessels travelling in parallel in the Strait at any given time. Also, when a vessel arrives at a station it may see another vessel passing the vessel in front of it. In this case, if the arriving vessel cannot reach the following station before the vessel in front of it, then it just follows it. If it can, then it checks to see if it can overtake the vessel passing the first one. If the arriving vessel can pass the vessel overtaking the first one, then it drops its speed to match with the first one and follows it in the regular lane. If not, it follows the overtaking vessel in the passing lane and passes the first one.

Furthermore, we assume that vessels do not stop for loading and unloading within the Strait and the local traffic does not interfere with the transit vessel traffic.

11.5.5 Simulation Model

The simulation model replicates operations at the Strait including various cargo and vessel types, their arrival patterns, and the operational details of the VTS using the Arena simulation software. The rules and regulations, pilotage and tugboat services are also considered in the model. The meteorological and geographical conditions such as fog, surface currents, and storms, are modelled in the simulation model using separate sub-models in the Arena. The model also includes an animation component as shown in Fig. 11.8. It shows the transit vessel movements in the Strait and the anchoring area as well as the waiting queues and some waiting time and transit time statistics.

11.5.6 Measures of Performance

The objective of the simulation model is to understand delays and bottlenecks in the Strait. This will be done through the analysis of various measures of performance estimated using the model. These performance measures include:

- Average transit times
- Average waiting times per class of vessels at the entrances
- Average waiting times due to closures
- Average waiting times of dangerous cargo vessels due to regulations
- Pilot utilisation
- Tugboat utilisation
- Number of vessels overtaking at certain points in the Strait
- Number of vessels navigating through certain sections of the Strait at any point in time (vessel density)

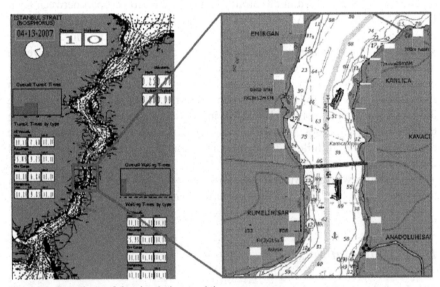

Fig. 11.8 Snapshots of the simulation model

Table 11.1 Total number of vessels per year

| Direction | Simulation | | | | 2005 data | |
	Average	Half width (95% CI)	Min.	Max.	Total	Relative error (%)
NB	26,940.32	1,007.94	26,660.60	27,083.80	27,402.00	1.68
SB	27,772.04	75.824	27,449.20	27,977.80	27,388.00	1.40
Total	54,712.36	123.256	54,109.80	54,971.80	54,790.00	0.14

11.5.7 Validation

In order to validate the simulation model, we have compared its output statistics with the data from the operation in 2005 provided by the VTS. The following simulation results are obtained from 10 replications, each 10 years long. The actual number of vessels that passed through the Strait in 2005 and the mean of the total number of vessels that passed per year in the simulation are given in Table 11.1, where NB and SB stand for northbound and southbound, respectively.

The average transit times of vessels obtained from both simulation and the 2005 data are shown in Table 11.2. The results seem to be quite accurate given the variations in the 2005 data.

The average waiting times of different types of vessels obtained from the simulation runs and the 2005 data have discrepancies. The average waiting times of dangerous cargo vessels obtained from the simulation are very close to the real data. However, the average waiting times of all vessels and tankers are significantly lower than the ones in the 2005 data. On the other hand, the values for the passenger vessels obtained from the simulation are considerably higher than the real data.

One reason for the lower waiting times in the model is the traffic interruptions due to construction of tunnels in Strait, which are not included in the simulation model. Therefore, the waiting times in the 2005 data are higher than the ones observed in the model. Another reason may be the lack of a standard scheduling algorithm used by the operators, which also explains higher waiting times for passenger vessels in the simulation. As stated earlier, a scheduling algorithm has been developed and incorporated in the simulation model. Therefore, the model does not take into account possible spontaneous decisions of the operator in charge. Also, lower waiting times obtained from the simulation are promising in terms of the effectiveness of the scheduling algorithm that we have developed.

Based on the comparison of the results obtained from the simulation and the real data collected in 2005, the total number of vessels passing through the Strait, the average transit times, and the average waiting times seem reasonable. Therefore, the simulation model is considered to be adequately representing the general behaviour and trends of the real system.

Table 11.2 Average transit times (minutes) of different types of vessels

Vessel type	Simulation		2005 data			
	Average	Half width (95% CI)	Average	Min.	Max.	Relative error (%)
All vessels	97.3952	0.02	99.1544	51.0167	9,271.37	1.77
General cargo	98.6468	0.01	98.9569	51.1667	2,268.82	0.31
General cargo NB	98.8776	0.03	101.1077	52.75	1,887.18	2.21
General cargo SB	98.4231	0.02	96.8061	51.1667	2,268.82	1.67
Dangerous cargo	86.8614	0.15	87.3001	59.8667	701.8833	0.50
Dangerous cargo NB	83.9690	0.19	89.546	59.8667	701.8833	6.23
Dangerous cargo NB	89.5104	0.19	85.0542	61.25	149.0833	5.24
LNG-LPG	88.2230	0.14	93.4553	63.9667	176.8833	5.60
LNG-LPG NB	87.4381	0.16	94.2355	63.9667	176.8833	7.21
LNG-LPG SB	96.4023	0.7	88.949	64.75	114.75	8.38
Tanker	92.8396	0.06	96.1582	51.0167	9,271.37	3.45
Tanker NB	93.0770	0.12	99.2164	51.2833	1,528.40	6.19
Tanker SB	92.6324	0.07	93.0999	51.0167	9,271.37	0.50
Passenger	96.7084	0.26	100.1648	54.05	1,731.65	3.45
Passenger NB	96.5936	0.47	100.439	54.65	812.1167	3.83
Passenger SB	96.8238	0.32	99.8892	54.05	1,731.65	3.07

11.5.8 Analysis of System Behaviour

First, we investigate the effects of some of the system attributes of concern on the system performance through several scenarios. These attributes are:
- Vessel arrival rates
- Number of available pilots
- Number of available tugboats

The available number of pilots and tugboats are treated as a group. The system attributes and their values applied to five distinct scenarios are displayed in Table 11.3. The shaded values correspond to the Base Scenario, which represents the current conditions in the Strait. All the consequent scenarios obtained by changing the attribute values according to Table 11.4 are compared with the Base Scenario in the following section. Each scenario is run for 10 years. The results for two of these scenarios are given next. The outcomes for the rest of the scenarios are consistent with the results explained below. Scenarios are discussed in the following sections.

Table 11.3 Values of system attributes used in scenario analysis

Attributes	Values			
Arrival rate increase	0%	5%	10%	15%
No. of pilots, tugboats	12, 4	16, 6	20, 10	–
Time between Class D, E and P vessels	5 min	10 min	–	–

Table 11.4 Average waiting times (minutes) in Scenario 1 compared with Base Scenario

Vessel type	Base Scenario		Scenario 1		
	Average	Half width (95% CI)	Average	Half width (95% CI)	Increase in average (%)
All vessels	342.64	107.98	318.54	39.77	−7.03
General cargo	242.91	40.76	236.34	31.66	−2.70
General cargo NB	214.78	40.24	206.72	37.42	−3.75
General cargo SB	270.18	41.83	265.04	27.88	−1.90
Dangerous cargo	694.52	257.11	631.46	80.3	−9.08
Dangerous cargo NB	646.40	232.56	579.25	92.09	−10.39
Dangerous cargo SB	738.57	279.99	679.45	77.76	−8.00
LNG–LPG	1,243.29	797.35	1,067.88	162.88	−14.11
LNG–LPG NB	1,299.94	850.09	1,103.22	164.96	−15.13
LNG–LPG SB	655.41	233.45	673.98	167.86	2.83
Tanker	802.13	399.32	702.16	81.72	−12.46
Tanker NB	777.19	405.07	671.74	78.36	−13.57
Tanker SB	823.92	394.56	728.90	85.04	−11.53
Passenger	77.9315	10.07	75.6351	4.45	−2.95
Passenger NB	73.8581	11.61	70.6946	4.97	−4.28
Passenger SB	81.9032	9.26	80.6488	5.49	−1.53

11.5.8.1 Scenario 1

- Arrival rate increase = 0%
- Number of pilots, tugboats = 20, 10
- Time between Class D, E and P vessels = 10 min

The model has shown that the number of available pilots and tugboats does not have any effect on the transit times of the vessels. However, as seen in Table 11.4, the average waiting times decrease consistently when the number of available pilots and tugboats increases from 16 and 6 to 20 and 10, respectively. The increase in the available number of resources affects mostly the average waiting time of the northbound general cargo and dangerous cargo vessels. The change affects the average waiting time of the southbound passenger vessels the least, which is consistent with the system policies since passenger vessels may get extra pilots and tugboats from the Port of Istanbul in any case.

Finally, the pilot and tugboat utilisations are given in Table 11.5. According to these figures, the pilot utilisation and the tugboat utilisation are around 30% and 2%, respectively for the Base Scenario. Although there is no related data for comparison, these values seem reasonable, taking into consideration the pilot and tugboat request percentages and the expected number of vessels in the Strait. On the other hand, the resource utilisations come down to approximately 20% and 1% for Scenario 1. Therefore, we observe that the pilot and tugboat utilisations decrease 25% and 40%, respectively in Scenario 1 compared with Base Scenario.

11.5.8.2 Scenario 2

- Arrival rate increase = 10%
- Number of pilots, tugboats = 16, 6
- Time between Class D, E and P vessels = 10 min

As expected, the arrival rates do not have any effect on the transit times of the vessels. These results are consistent with the system structure since the time a vessel spends in the Strait is not dependent on the vessel inter-arrival; it is affected by its speed, the traffic density and the current and visibility conditions in the Strait. However, as seen in Table 11.6, a 10% increase in the arrival rates leads to an almost 270% increase in the average waiting times of vessels in general. Specifically, the increase in the number of vessel arrivals affects the average waiting time of the dangerous cargo vessels the most.

Finally, the pilot and tugboat utilisations are given in Table 11.7. According to these figures, the pilot utilisation increases by almost 15% as the number of vessel arrivals increases by 10%. The increase is around 12% and 15% for the southbound

Table 11.5 Resource utilisations in Scenario 1 compared with Base Scenario

	Base Scenario		Scenario 1		
Resource	Average	Half width (95% CI)	Average	Half width (95% CI)	Increase in average (%)
Pilot NB	0.3077	0.0042	0.2307	0.0035	−25.01
Pilot SB	0.2838	0.0033	0.2156	0.0023	−24.02
Tugboat NB	0.0120	0.0004	0.0069	0.0002	−42.81
Tugboat SB	0.0312	0.0007	0.0186	0.0005	−40.28

Table 11.6 Average waiting times (minutes) in Scenario 2 compared with Base Scenario

	Base Scenario		Scenario 2		
Vessel type	Average	Half width (95% CI)	Average	Half width (95% CI)	Increase in average (%)
All vessels	342.64	107.98	1,258.23	1,006.17	267.22
General cargo	242.91	40.76	873.67	658.14	259.67
General cargo NB	214.78	40.24	791.11	631.57	268.34
General cargo SB	270.18	41.83	953.93	684.87	253.07
Dangerous cargo	694.52	257.11	3,364.60	2,945.53	384.45
Dangerous cargo NB	646.40	232.56	3,430.84	3,060.95	430.76
Dangerous cargo SB	738.57	279.99	3,300.18	2,830.15	346.83
LNG-LPG	1,243.29	797.35	2,641.21	1,919.45	112.44
LNG-LPG NB	1,299.94	850.09	2,789.22	2,055.30	114.57
Tanker	802.13	399.32	3,122.03	2,681.13	289.22
Tanker NB	777.19	405.07	2,953.30	2,533.63	280.00
Tanker SB	823.92	394.56	3,271.03	2,814.10	297.01
Passenger	77.9315	10.07	87.0963	7.23	11.76
Passenger NB	73.8581	11.61	82.0913	5.41	11.15
Passenger SB	81.9032	9.26	92.1096	9.23	12.46

Table 11.7 Resource utilisations in Scenario 2 compared with Base Scenario

	Base Scenario		Scenario 2		
Resource	Average	Half width (95% CI)	Average	Half width (95% CI)	Increase in average (%)
Pilot NB	0.3049	0.00	0.3409	0.00	11.81
Pilot SB	0.3009	0.00	0.3456	0.00	14.86
Tugboat NB	0.0112	0.00	0.0125	0.00	11.61
Tugboat SB	0.0152	0.00	0.0174	0.00	14.47

and northbound tugboat utilisations, respectively. The increase in the northbound tugboat utilisation is more significant than the southbound tugboat utilisation due to the higher percentage of tugboats requested by the northbound vessels.

11.6 Conclusions

In this case study, the transit vessel traffic system in the Strait of Istanbul is modelled using the Arena simulation tool to mimic and study the operational aspects of the maritime traffic and the risks involved in it. The case presented here has covered only the operational aspects. The model assumptions and details include cargo and vessel types, regulations for vessel movements, and weather and water conditions, as well as the use of various resources such as pilots and tug boats. Additionally, the model incorporates the current vessel scheduling practices via a scheduling algorithm. This algorithm is developed through discussions with the Turkish Straits VTS to mimic their decisions on sequencing vessel entrances as well as coordinating vessel traffic in both directions. Furthermore, a scenario analysis is performed to evaluate the impact of several parameters on the system performance.

The model that is briefly presented here is a part of a major initiative to identify risks generated by the vessel traffic in the Strait to the town of Istanbul, its inhabitants and environment, measure them on the basis of international standards and produce policies to mitigate them.

11.7 Questions

1. Why is simulating the vessel traffic in the Strait of Istanbul important?
2. What is the ultimate goal in a waterway simulation? How would one use it?
3. What type of input data does one need to simulate waterway traffic?
4. What measures of performance should be obtained in waterway traffic simulations?
5. What types of scenarios should one look at in waterway simulations?
6. How do you model vessel movements in waterway simulations?
7. Develop a conceptual model to manage the daily vessel traffic in the Strait of Istanbul. This is not a simple task; however, think in terms of simple arguments.
8. Propose a simple queuing model to estimate the average vessel waiting time at waterway entrances.

Acknowledgements This work is in part funded by the Laboratory for Port Security at Rutgers University, NSF Grant Number INT-0423262, and TUBITAK, The Scientific and Technological Research Council of Turkey through the Research Project 104Y207 and BAP, Scientific Research Projects Fund of Boğaziçi University through the Research Project 07M104. We are indebted to the Turkish VTS Authority for providing us with the data on the vessel traffic in the Strait of Istanbul.

References

[1] Nitzov B (1998) The Bosphorus: Oil through needle's eye? Institute for Energy Economics and Policy, Sarkeys Energy Center of the University of Oklahoma, OK

[2] Regulations (1994) Maritime traffic regulations for the Turkish Straits and the Marmara Region

[3] EIA (2005) World oil transit chokepoints. US Energy Information Administration. http://www.eia.doe.gov/emeu/cabs/World_Oil_Transit_Chokepoints/Bosporus_Turkishstraits.html. Accessed November 2006

[4] Köse E, Başar E, Demirci E et al (2003) Simulation of marine traffic in Strait of Istanbul. Simul Model Pract Theory 11(7–8):597–608

[5] Merrick JRW, Van Dorp JR, Blackford JP et al (2003) A traffic density analysis of proposed ferry service expansion in San Francisco bay using a maritime simulation model. Reliab Eng Syst Saf 81(2):119–132

[6] Özbaş B (2005) Simulation of maritime transit traffic in the Istanbul Channel. Department of Industrial Engineering, Boğaziçi University

[7] Almaz AO (2006) Investigation of the transit maritime traffic in the Strait of Istanbul through simulation modeling and scenario analysis. Department of Industrial Engineering, Boğaziçi University

[8] Almaz AO, Or I, Özbaş B (2006) Simulation of maritime transit traffic in the Istanbul channel. In: Proceedings 20th European conference on modelling and simulation, ECMS 2006

Chapter 12
Airport Logistics Operations

Miquel Àngel Piera Eroles, J. José Ramos and E. Robayna

Abstract In recent years, air traffic has increased dramatically while airport capacity has remained stagnant. This has resulted in congestion problems which degrade the performance of the air traffic control system and cause excessive costs. Despite recent technological advances in the airport airside area, some procedures and operational rules in the landside area are years behind airside capability. In this chapter, a discrete-event system view of airport operations is introduced. The main aspects of delay propagation due to a lack of coordination policies will be illustrated using an Arena$^\copyright$ simulation model.

12.1 Introduction

In the air traffic management (ATM) context, the terminal manoeuvring area (TMA) is the most complex subsystem due to the dynamics of the aircraft movements in the airside (conflict-free trajectories) and the scheduling of the airport infrastructure (runway, taxiway, parking, gates) together with the services (ground handling segment). Nowadays, TMAs are the main bottleneck to supporting the future expected increase in air traffic flow capacity. Furthermore, the TMAs are the areas that urgently require operational efficiency improvements in the airport airside and landside operations.

Miquel Àngel Piera Eroles and Juan José Ramos
Universitat Autònoma de Barcelona, Spain
MiquelAngel.Piera@uab.es, juanjose.ramos@uab.cat

Ernesto Robayna
AENA, Palma de Mallorca Airport, Spain
erobayna@aena.es

Y. Merkuryev et al. (eds.), *Simulation-Based Case Studies in Logistics*
© Springer 2009

Airlines are constantly demanding a reduction in the waiting time at the end of the runway to take-off and in the holding trajectories for landing, which results in poor cost effectiveness due to excess fuel-burn and wastage of time. Thus, there is an economical and social motivation to focus extra research efforts in order to solve the congestion problems in the TMA. In fact, the pressing need to improve airport efficiency has also been confirmed by an analysis of Eurocontrol data, in which three-quarters of the delays longer than 15 minutes (with respect to the planned times) generated in airports are due to poor activity coordination.

There are several recent technological advances, such as new aircraft with greater fuel efficiency, huge air freighters, an expanding general aviation fleet, together with better navigation and surveillance technology (ADS-B, satellite navigation, GPS, etc.) that are paving the way to a competitive air transport system. Nevertheless, delays are still generated and propagated in most airports. Improving air transport KPIs requires not only addressing the technical aspects, but also the tactical and operational procedures that condition both operational effectiveness and economic practicality.

Solutions to this problem vary according to the planning horizon. Long-term considerations involve building new airports and additional runways. Medium-term approaches focus on ways to disperse traffic to less-busy airports through regulations, incentives, etc. Finally, short-term solutions aim to minimise the unavoidable delay costs under the current capacity and demand. This chapter will focus on the airport dynamics that belong to the latter category.

12.1.1 The Current System

Some of the early strategies developed to handle the above problems started by improving airport infrastructures, e.g. building additional runways, taxiways, or terminals [1] and increasing handling resources. Stand-alone solutions, like additional radar, or control tower extensions, were also established; however, most airport managers realised after a short while that oversizing infrastructure and updating technology were not synonymous with airport efficiency. Instead, airport taxes increased, delay propagation at the operational level remained, and the passenger service quality factor (SQF) did not improve proportionally to the increase in airport taxes.

Despite the fact that new functionality was introduced through different technological changes, such as replacing old aircraft with new ones, thus expanding new capabilities of efficient aircraft operating possibilities, the rate of introduction of new tactical and operational ground airport alternatives has hardly changed.

It should be noted that congestion problems become more serious when air traffic increases. Airport resources are often exhausted, working at their limit, while at the same time some resources are idle or oversaturated during certain time frames. Air transport market competition and a lack of partnerships between handling operators, together with unpredictable arrival/departure aircraft times, make a deeper knowledge about airport dynamics necessary in order to improve its ability to respond efficiently to any time deviation with respect to the proposed scheduling.

Airport operational activities should be understood as a highly coordinated evolutionary process which requires planned periodic changes rather than reactive changes in configuration, with a procedural collaboration between the different airport operators (see Sect. 12.2). Nowadays, ground-based ATM capability in some areas is years behind airside capability; many of the landside operational procedures are based on the extension of past and current operating practices. To define new operational procedures at fundamental levels, the roles and operating methods of all members that interact in an airport's operational activities must be identified and the propagation of the consequences of their decisions should be properly understood in order to be able to deal with a safe and economically viable system that provides benefits and allows for sustained growth.

In order to handle adequately airport decision support tools and adapted operational procedures for the operators, a deep knowledge of the interactions, as well as the quantity and quality of the different relationships between the operations performed by the different airport operators, is essential to properly address the on-ground problems. Airport operators must have the means (systems and procedures) to coordinate the different operational actions in order to hand over efficiently the control of each aircraft.

Simulation models can contribute immensely to a better understanding of all interactions and of the different consequences of any decision made by an airport operator, and can help design the improvements and procedures that can be most easily integrated into a continuously evolving transitional process involving systems with a wide range of capabilities.

12.2 Main Airport Subsystems

A high-level description of the main phase sequence through which an aeroplane must flow from its arrival to the TMA until its departure will be introduced in order to provide a better understanding of the airport decision variables and their impact on overall airport performance. This section includes background information on an airport's operational environment. It describes the main partners and their services with the aim of facilitating the understanding of the main interaction that should be modelled to improve overall airport efficiency.

Airport decision-making is primarily carried out by four main operators: airlines, ground handling segment, airport operations and air traffic controllers (ATCs). Airline operators and handling agents can sometimes be grouped as one entity, since handling agents are considered to be the representative of the aircraft operator and act on their behalf. In fact, at an airline's home base, all the activities are normally performed by the aircraft operator's own staff. At other airports, some of the activities are delegated (outsourced) to the handling agent.

Decisions are made with the objective of optimising airport capacity (use of available capacity through the maximisation of resource use, throughput, etc.). These are based on information (e.g. arrival and departure time estimates) that changes over time or has poor timeliness/accuracy. Such data is received from various sources (airlines, handlers, local ATCs, etc.) and constrained by several variables: the operating airline's schedule, aircraft type, destination country, passenger information, terminal and pier capacity, etc. Unfortunately, each operator runs its own information distribution system, collecting data from dedicated sources within its own domain, without cross-fertilisation of information between the different operators.

12.2.1 Airport Operators

Airport operations are responsible for the management and allocation of airport resources, such as the planning of stands and gates, check-in desks, baggage reclaim belts, apron management and security management. In certain airports, aircraft operators and/or ground handlers self-manage their allocated resources (stands, check-in desks, etc).

Despite the fact that a first estimate for stand and gate allocation for a specific aircraft can be computed when the flight takes off from the originating airport, gate assignment decisions are often made at the very last moment when the aircraft is landing, due to the lack of an accurate landing estimate (landing sequence is established 10–15 minutes before landing when the flight passes into the approach area), unexpected delays in taxi-in time, or limited knowledge about the pushback of the preceding aircraft at the gate.

A last-minute gate assignment could avoid these drawbacks; however, lead time differences between aircraft movement and passenger movement in the terminal platform limits this possibility. It should be noted that airport operation planners only update the original gate assignment if there is a significant delay in the off-block time (in principle, more than 15 minutes). This gap can cause some aircraft to be directed to remote points while there are free contact points.

12.2.2 Air Traffic Controllers

The ATC tower is in charge of aircraft operating in the airport's manoeuvring area and within the airspace around the airport (holding trajectories). With regard to airport logistics, the ATC can be seen as the boundary conditions of the airside workload at the airport, by establishing the arrival and departure sequence based on the traffic at the holding points.

Other decisions made by the ATC are: issuing clearance for pushback and taxi, guiding aircraft from the parking position to the holding point for outbound flights and from the runway exit point to the apron entry point for inbound flights.

Any change with respect to the scheduled times is propagated as delays or resource idleness to the other airport operators.

12.2.3 *Airlines and Ground Handling Segment*

Ground handlers provide services to both aircraft and passengers. Some of the tasks associated with passenger services are:
- Lounges and VIP services
- Passenger assistance
- Check-in, gate and transit
- Ticketing

Some of the tasks to be coordinated in relation to aircraft services are:
- Baggage transportation
- Aircraft loading and unloading
- Ramp support
- Pushback
- De-icing
- Operation control
- Load planning
- Supervision
- Ground equipment maintenance

Figure 12.1 illustrates different ground handling operations that should be properly coordinated to avoid delay propagation to the other airport operators.

Aircraft operators (AOs) are responsible for complying with their assigned slot. Most AOs use conservative models to estimate the taxiing period in order to be sure that the aircraft will be ready for start-up in sufficient time, which leads to a situation where many aircraft are often waiting at the end of the runway, resulting in very poor KPIs in relation to efficiency, environment and cost effectiveness, due to excess fuel-burn and wastage of time.

The taxi period includes not only the time to taxi from the parking area to the end of the runway (a default value for each particular airport), but also the waiting time at the runway, which in turn depends on various data, such as the number of arriving flights and departing flights.

Fig. 12.1 Ground handling operations

12.3   Collaborative Decision Approach Benefits

Despite the fact that there is an unavoidable cost due to changes in weather conditions (e.g. visibility, wind) which can justify a drop by half or even more in airport capacity due to bad weather conditions, there is a considerable amount of delay generated at the airport and propagated through the operational activities (http://www.euro-cdm.org) resulting from poor coordination of the operations [2].

Nowadays, from a functional point of view, most airport activities are considered and tackled in an independent way by different departments. Under the present operational situation, any perturbation can be easily propagated through the airport, affecting passenger service quality factors and airline company costs. Some examples of problems that could be mitigated with better knowledge of airport dynamics are:

- Gates are allocated to flights based on their scheduled arrival times. Lack of exact arrival time information when flights are late can result in empty contact stands while other aircraft are parked at remote stands. This increases turnaround times because handling resources located at the contact stand must be moved to remote stands, or remain idle near the contact stand.
- Changes in the landing sequence made by ATCs may allow certain aircraft to arrive earlier to the parking area. If handling or ground crew are not ready to handle the aircraft, the disembarking operation will be delayed, thus decreasing SQF because passengers will be forced to wait, and increasing airline costs because terminal occupancy will be higher.
- Handling resources (e.g. pushback trucks) are not always in place under the right aircraft because the staff is unaware of the departure sequence or pushback sequence.

A single delay in a certain operation can be easily propagated through all the airport subsystems. It is easy to notice that in order to avoid idleness in handling resources, handling operations should be scheduled to saturate workers and resources while providing a timely service. In this context, a delay in the start of the pushback operation will cause a delay in the freedom of the truck, which will force a delay in attending to the next pushback operation.

The design and proper application of new operational procedures that could take in consideration the state of the airport at any moment will provide better SQF to passengers and can propagate benefits to the different airport operators [1].

Some benefits provided to ground handling operators are:

- Improved pushback productivity thanks to better use of staff and reduced inactive time due to inefficiencies (e.g. less time wasted by ground vehicles)
- Reduction of (indirect) operating costs as a result of a reduction in delays
- Knowledge of the precise status of arriving aircraft well in advance that will optimise the handling of flights

Some benefits provided to airline operators are:

- Pre-departure sequence can be optimised, better ground movement and more efficient take-off order, less idling on the ground.

- More capacity maintained during adverse conditions and the return to normal conditions can be faster. Both can result in major cost savings.
- Optimisation of gate utilisation and other ground resources. The effects of late incoming or departing flights and missed connections can be reduced.
- Greater predictability leads to greater use of staff resources since rosters can be organised to meet demand. As a result, crew management costs can be reduced.

Some benefits provided to the ATCs are:
- A collaborative pre-departure sequence enables ATCs to take user preferences into account.
- Accurate taxi times increase the accuracy of the calculations in which taxi times are used, improving predictability (benefit to all partners).
- Constant work load, preventing controllers from becoming fatigued due to work overload.

Some benefits provided to airport operations are:
- Reduced delays and hence greater predictability leads to a greater use of staff resources since rosters can be organised to meet demand. As a result, staff employment costs can be reduced.
- Better information related to the departure and arrival sequence can result in a significant improvement in the planning capability for further operations and also allows better quality information to be dispatched to relevant partners (e.g. passengers and handling agents).
- Having knowledge about the departure sequence should improve the allocation of stands and gates.

12.4   A Discrete-Event System Approach

In this context, it is important to view the operations from the airport perspective. For the airport, the flight has three phases: an inbound phase, a ground phase and an outbound phase (see Fig. 12.2). A delayed inbound flight has an impact on the ground phase, but also on the outbound phase of the flight with the same airframe, on the crew and on the flights carrying connecting passengers.

To avoid delay propagation, a deep knowledge about all the events that take place and their interactions in each phase is important. Thus, by considering the ground phase, the turn-around, landing, take-off and taxiing operations can be formalised as a set of inter-related events which, properly coordinated, will satisfy the aircraft operative needs under certain SQF. With a proper model specification

Fig. 12.2 The three flight phases

considering its interactions with inbound and outbound phases, it will be possible to optimise operation efficiency through the proper management of airport resources (e.g. airport slots, stands and gates, check-in desks and baggage belts), considering the dynamics and costs of the passenger and aircraft operations.

In the particular case of turn-around operations, it is easy to understand the system dynamics from a discrete-event system approach, in which each operation has a certain number of pre-conditions, a duration time estimation, and a set of post-conditions (changes in the state of airport information). Figure 12.3 illustrates the different handling resources that should be properly coordinated to provide an efficient service to the aircraft.

To improve ground handling performance, a discrete-event simulation model that could provide resource planning and staff allocation closer to scheduled times would eliminate the need to plan additional time buffers for staff in order to cover delayed outbound flights. Some indirect consequences of improving ground handling efficiency would be higher productivity, and thus higher revenue or a reduction/elimination of operating costs, e.g. greater use of resources leads to a reduction in current operating costs (like those generated by resource allocation conflicts) and the elimination of future operating costs (less need to hire staff and buy equipment thanks to greater use of existing resources). Poor rescheduling when flights are delayed increases the volatility of the resources needed. As a result, the redundancy required is impacted, as is the cost of providing the service.

A coloured Petri net model describing the sequence of the turn-around operations was developed to tackle the ground handling operations from a logistics point of view. In this model, it is possible to apply different scheduling and planning policies in order to provide a proper answer considering the well-known '7 Rs rule': 'Ensure

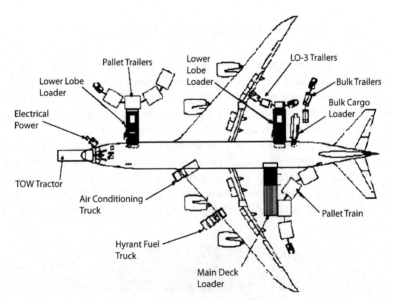

Fig. 12.3 Turn-around aircraft locations for ground handling

the availability of the Right product, in the Right quantity and Right condition, at the Right place, at the Right time, for the Right customer, at the Right cost'.

12.5 Palma de Mallorca Airport: Check-In Assignment Sensibility

Palma de Mallorca Airport (PMA; http://www.aena.es) is considered the third busiest Spanish airport [3] regarding the flow of passengers/year (23,228,879 passengers in 2007) and the number of aeronautical operations (197,384 movements in 2007).

The main infrastructure characteristics are:

- Two runways that can be operated independently: both can be configured for landing operations or take-off (06R has some restrictions due to environmental measures).
- 28 contact points and 42 gates for remote points distributed in four different terminals.
- 204 check-in points.
- A maximum airside capacity of 60 movements/hour: 32 arrivals/hour, 30 departures/hour.
- A maximum landside capacity of 6,000 pax/h (outbound passengers), and 6,300 pax/h (inbound passengers) (5,600 EU pax/h and 700 non-EU pax/h).

Figure 12.4 illustrates the runway and terminal configuration at PMA.

Fig. 12.4 Palma de Mallorca runway and terminal configuration

To improve the passenger and airline quality factors, PMA has recently designed a new functional area called the 'Production Department'. This department works to ensure proper passenger, baggage and aircraft synchronisation during the boarding operation, considering the quality service and security levels defined by the airport and the present standards. The lower part of Fig. 12.5 shows the classical PMA model approach used to address the airport operational activities. The upper part of the same figure illustrates the functions of the new Production Department.

As can be seen in Fig. 12.5, the new Department seeks to coordinate the planning of the Airside Operations, Terminal Operations, Security & Safety and Infrastructure and Information Technology Systems Departments, in order to:

- Monitor and supervise the state of the airport at any moment.
- Coordinate the best actions to be implemented in each department.

12.5.1 Delay Propagation in the Passenger Flow Area

The check-in processes at PMA are grouped into two primary areas at the main entrance building (at floor level). They are distributed in six parallel blocks with 32 counters in each (see Fig. 12.6).

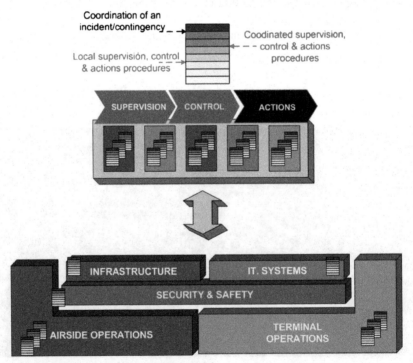

Fig. 12.5 Real-time airport management model

The check-in assignment is highly flexible, so the operational assignment can change daily according to traffic demand, traffic typology (i.e. regional, charter or conventional airlines, individual or block operations) and commercial aspects.

Before passengers can access the terminal area, once they have checked-in, they must pass through the security filters, which are placed at both sides on the second floor (see Fig. 12.7). Security is a very sensitive process which requires intensive human and technical resources and can drastically influence the time required by passengers to move from the check-in area to the gate. It can also influence turn-around time. Thus, to avoid extra delays, it is important to rearrange the number of open security filters on each side in advance (planning policy) or to redirect the passenger traffic to the opposite security area in order to balance the queues (reactive policy).

Since passengers will choose the security area closest to their check-in area, the check-in assignment model should consider the workload estimations in each security area to improve security assignment planning. Figure 12.8 shows the estimated

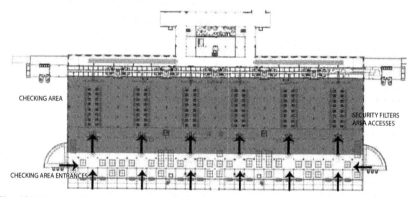

Fig. 12.6 Layout distribution of the check-in counters at Palma de Mallorca Airport

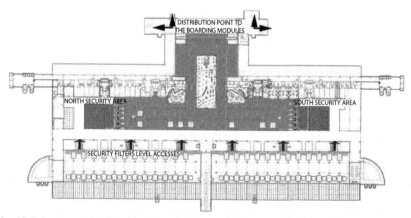

Fig. 12.7 Layout distribution of the security area at Palma de Mallorca Airport

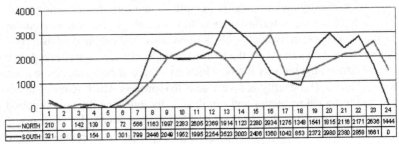

Fig. 12.8 Estimation of outbound passenger distribution at PMA

number of outbound passengers distributed according to a developed model in the north and south security areas to avoid long queues in one area while there is idleness in the other.

It is easy to note that in the event that the check-in process in the 1.00–2.00 p.m. interval is delayed due to a perturbation (Automatic Baggage Management System off, inexperienced personnel at the check-in counters, bus passenger arrival delayed due to city traffic jump, etc.), the security process will be overlapped with the workload estimated for the 3.00–4.00 p.m. timeframe at the opposite side (north security area). Something similar would happen if the check-in process during the 3.00–4.00 p.m. interval was advanced (earliness situation). The short time to react to security over-saturation will be propagated to stands, gates, boarding operations, handling requirements and, unfortunately, a departure delay.

The cause–effect analysis of a model considering the specification of the different events that interact through the different airport processes has contributed to the design and justification of new alternative procedures, such as:

- 'Last minute' check-in counters and security filters that can be used as a decision variable to avoid delay penalisations to certain turn-around processes (especially those with a shorter turn-around time).
- Selectively slowing down certain check-in processes to avoid an unbalanced security workload and to increase the time to redirect the flow of passengers.
- Specific check-in processes that can be performed at the terminal gate, thus uncoupling certain infrastructure and security operations.

12.5.2 Delay Propagation in the Passenger Transfer

Airlines try to concentrate arrivals and departures within a narrow timeframe due to commercial motivations, crew roster costs and resource minimisation. At PMA, the co-existence of the Air Berlin hub and the German and UK flight banks throughout the day generates emergent dynamics due to transfer connections when some flights arrive to PMA delayed. Figure 12.9 shows the three typical banks of Air Berlin on two different days. More than 20 aircraft arrive to PMA from different German

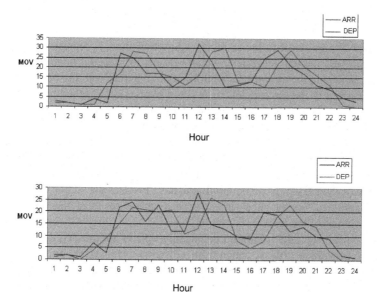

Fig. 12.9 Palma de Mallorca Airport: Air Berlin banks

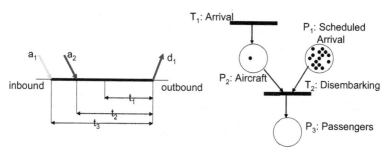

Fig. 12.10 Minimum connecting time and its PN model

origins within the timeframe of 1 hour, and they depart 90 minutes after their arrival to different Spanish airport destinations, mixing passengers from different aircraft (transfers).

The *minimum connecting time* is defined as the gap of time between the arrival of the last aircraft and the departure of the first aircraft. This is a critical factor that constrains turn-around time. Figure 12.10 illustrates the minimum connecting time concept, in which:

- a_1 is the planned arrival time of the last bank aircraft.
- a_2 is the arrival time of the last bank aircraft.
- d_1 is the departure time of the first bank aircraft.
- t_1 is the minimum connecting time.
- t_2 is known as the effective transfer time.
- t_3 is known as the scheduled transfer time.

On the right-hand side of Fig. 12.10, the conceptual model of the minimum connecting time has been represented in Petri net formalism. Transition T_1 represents the arrival of an aircraft to the in-block, while transition T_2 represents the disembarking process, which generates as many tokens to place on node P_3 as passengers that arrived in the aircraft. This information, together with the next connecting flight is carried in the aircraft token. A guard expression attached to transition T_2 compares the arrival time with the expected arrival time, to activate new procedures in case the minimum connecting time is not preserved.

Based on the final destination dispersion level of the passengers of each arrival flight and the distance between the gates assigned to each aircraft, different key performance indicator values of the available resources can be obtained. The fact that some of the passenger outbound flights have PMA as the source airport should also be examined. So, the check-in and security processes must also be considered in the dynamics of the transfer operation.

Sometimes, when the minimum connecting time is not preserved, all departing flights are delayed in a block, which from a discrete-event system point of view can be interpreted as a single event (an inbound delayed flight) that can freeze the firing of several events (departing flights). Coloured Petri net formalism makes it possible to represent, in an easy-to-understand way, this type of cause–effect relationship and evaluates new procedures to minimise delay propagation. Furthermore, most of the PMA departing flights in a first bank will come back to PMA during the course of the day (see Fig. 12.11). So, when departing flights are delayed waiting on the arrival of an aircraft, this delay will be propagated to other airports, and will affect PMA again with later delays.

12.6   Delay Propagation Simulation Model for Pushback Operations

Management of gate operations is a key activity at airports. Aircraft are assigned to terminal gates or ramp positions for the duration of a time period during which passengers and aircraft are processed. Amongst the three flight phases at the airport, the predictability at the second and third is not good enough: ground and outbound

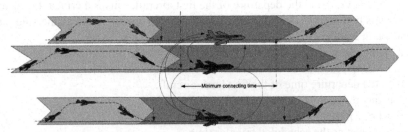

Fig. 12.11 Delay propagation throughout the day at PMA due to connecting flights

phases. Some statistics by Eurocontrol show that more than 22% of air transport delays are longer than 15 minutes (vs. schedule) and three-quarters of them are due to ground processes. As discussed before, delays at any of the flight airport phases have undesired consequences both upstream and downstream. Therefore improving the operation planning and management of these phases is important.

To illustrate the benefits of a simulation model that could consider the delays on the scheduled operations, the pushback operations will be considered. The example considers an airport where 30 departure operations have to be completed during a peak hour. The time elapsed between departure from the origin airport gate (push-back/out-block time: OBT) and wheels off (take-off time: TOT) is known as the taxi-out time. The model will represent the main operations of the out-block process. Usually, pushback trucks spend 15 minutes as an average with one aircraft while the actual process is only 5 minutes – so there is a good potential for improvement, maybe up to one-third of the time could be saved. A direct consequence of planning a more efficient operation is that fewer truck resources will be needed for a given number of operations. As an indirect consequence, aircraft delays caused by the lack of a pushback truck (which may be idling somewhere else and could not be repositioned) could be decreased and, therefore, an improvement of the SQF can be achieved as well.

The main purpose of the simulation model is to analyse different scenarios in which the performance of the system according to two different strategies can be compared: increasing the number of resources and introducing information to enable collaborative decision making. For simplicity reasons, a basic first-come first-served (FCFS) policy is adopted to assign trucks to pushback operations.

On the basis of deterministic information about the set-up time and lead time of the pushback operation (5 min will be assumed in both cases), it is determined that five trucks are needed to perform 30 operations per hour. The graph on the left of Fig. 12.12 shows a feasible schedule for this case. Gate and resources availability and times of arrivals/departures (as given by an estimated time) can change during the course of the planning horizon due to operational contingencies (for example, congestion, lack of capacity, air traffic control). In a realistic scenario, such a theoretical schedule will never work. Queuing theory can be also used in order to take into account some of the stochastic aspects of the system. Still, such models can hardly capture all the events which can deteriorate the system perfor-

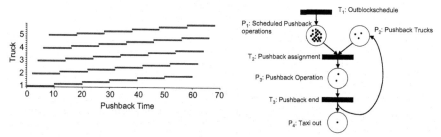

Fig. 12.12 Deterministic pushback scheduling and PN model of the pushback process

mance. A discrete-event simulation model can represent the stochastic behaviour and all the relevant events of the system, enabling an in-depth analysis of realistic scenarios.

On the right of Fig. 12.12 the conceptual model of the pushback process has been represented in Petri net formalism. Transition T_1 represents the schedule of an aircraft for the out-block. The place node P_1 represents all the aircraft scheduled for out-block (they are waiting for their pushback time and/or for the pushback truck become available). The place node P_2 represents the trucks which are ready for operation. The transition T_2 represents the assignment of an available truck to a scheduled aircraft whose out-block time has arrived. A guard expression attached to transition T_2 compares the current time with respect to the nearest expected out-block time in order to implement the FCFS assignment policy or to reschedule a new OBT if no truck is available. The place node P_3 represents the aircraft during pushback operation while the place node P_4 represents the aircraft at the taxi-out operation (this process is not modelled). Transition T_3 represents the end of the pushback operation and releases the pushback truck.

This model aims to illustrate the use of simulation as a means to analyse the system performance. It does not pretend to be an optimisation approach. The model has been implemented with the Arena simulation tool. In order to make the experimental results more comprehensive, only the OBT is modelled as a random variable.

No information about actual OBT (the instant when the aircraft becomes ready for the pushback operation) is considered in the first simulation scenario. Hence, the FCFS policy is applied over the estimated out-block time (EOBT). In this case, an

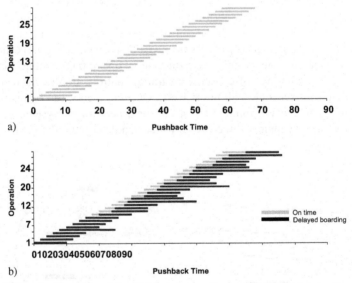

a)

b)

Fig. 12.13 Nominal operation (**a**) and delay propagation in the pushback operations (**b**)

available truck is assigned to the aircraft as soon as the EOBT is reached. If a delay (due to, for instance, the boarding process) appears then the assigned truck will be idle until OBT arrives, so its next assignments will be also delayed. The graph at the top of Fig. 12.13 shows the operation scheduling when no delay on EOBT appears. The graph at the bottom of Fig. 12.13 shows a more realistic scenario where OBT is moved ahead or delayed over EOBT. Five trucks are used in both cases. When a first delay occurs, it propagates over time showing an additive behaviour as new delay appears, since there is no available information about the possible earliness of some operations. Finally, 50% of flights show a delay greater than 1 minute in spite of only seven having a delay over the EOBT and eight being ready for out-block before their EOBT.

An obvious solution is to increase the number of resources. However, there are several limits: cost (evident), technical (constraints posed by the Airport Manager) and truck idleness (more trucks will not solve the problem of trucks assigned to a unready aircraft).

A second simulation model is set up in order to represent a scenario where the up-to-date information about the aircraft readiness for out-block is available, so CDM (collaborative decision making) is enabled. An FCFS policy is also applied but, in this case, using the actual OBT instead of the EOBT. Therefore, advantage of aircraft earliness can be gained. Both models are simulated with five to eight trucks in order to compare some illustrative indicators. The first significant measure is the flight idleness (elapsed time between aircraft's readiness and the initiation of out-block). As can be seen in Fig. 12.14, flight idleness is twofold without real-time information (RTI) and drops drastically when information sharing between airport operators is supported. The truck idleness (computed as the time elapsed between truck assignments and out-block initiation) is null with RTI, which is obvious since trucks are assigned as aircraft become ready. However, it increases without RTI as the number of trucks increase. It seems to indicate that the FCFS policy is not suitable when using EOBT. The graphs at the bottom show the absolute delay at each operation. It can be seen that better performance is achieved by using RTI. An also interesting figure, not included in the graphs, is the percentage of delayed aircraft. With five trucks, percentages are 50% (without RTI) and 27% with RTI. With six trucks, percentages drop to 10% and 7% respectively.

Finally, Fig. 12.15 shows the usage ratio of a truck. Without RTI and five trucks, the ratio is over 100% which means that the 30 pushback cannot be dispatched within 1 hour. Furthermore, the usage ratio stays very high even when the number of trucks increases (usage includes idle time since the resource is not available meanwhile). That makes the system very sensitive to faults. Once again, with RTI the usage ratio is 83% or less, except for the case of five trucks where the demand equals the capacity.

An interesting non-trivial question emerges: what is the most important for the best performance, an oversized set of resources or a proper information system enabling collaborative decision between different airport operators?

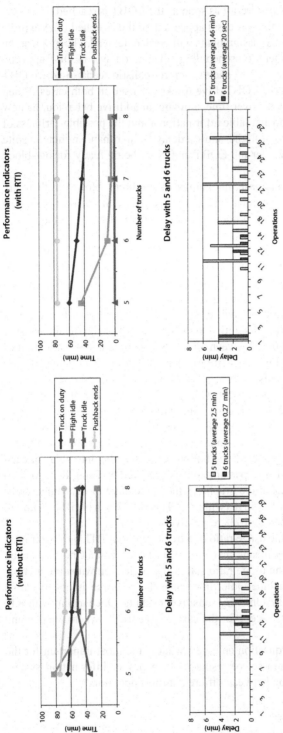

Fig. 12.14 Comparison of some performance indicators without and with real-time information

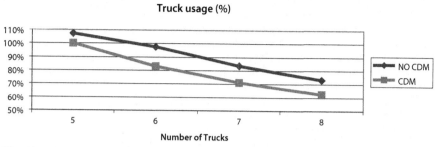

Fig. 12.15 Truck usage ratio

12.7 Conclusions

A discrete-event system view of certain cause–effect interactions in the main airport operations has been presented. Delay generation and propagation through different airport areas due to the poor coordination of interacting operations has been illustrated by means of real examples.

The use of planning and scheduling policies developed in the logistics area can considerably help in the understanding and improvement of the overall performance of airport operations, while providing benefits to all airport agents.

A simulation model to improve the productivity of pushback handling resources preserving service quality factors for airline companies has been developed to justify the advantages of using simulation technologies. These technologies can contribute to a deeper knowledge of airport dynamics and help design operational procedures which will mitigate perturbations.

12.8 Questions

1. What are the main consequences of the late arrival of an aircraft?
2. What are the main aspects that provoke uncertainty in pushback truck scheduling?
3. Why can't gate assignments be resolved at the last minute once the aircraft has landed?
4. What are the main aspects of airport flexibility that limit the use of classical optimisation techniques in order to deal with optimal scheduling policies?

Acknowledgment This work is partly funded by the Science and Innovation Ministry of the Spanish Government, 'Discrete-Event Simulation Platform to improve the flexible coordination of land/air side operations in the Terminal Manoeuvring Area (TMA) at a commercial airport', CICYT Spanish programme TRA2008-05266/TAIR.

References

[1] EATMP Information Centre (2003) Eurocontrol air traffic management strategy for the years
     2000+, vols 1, 2 (Eurocontrol ATM 2000+). Eurocontrol, Brussels
[2] Jasselin P, Sureda-Perez S et al (2001) A-CDM-D Final Report. TEN 45601
[3] Airport and traffic data. AENA. http://www.aena.es. Accessed 21 August 2008

Subject Index